A History of Telecommunications

A History of
Telecommunications

JOHN TYSOE
AND
ALAN KNOTT-CRAIG

BOOK**STORM**

For Olivia, Barnaby, Henry and Teddy
– John Tysoe

For Mom and Dad
– Alan Knott-Craig

'We do not see that this device will be ever capable of sending recognizable speech over a distance of several miles. Messrs Hubbard and Bell want to install one of their "telephone devices" in every city. The idea is idiotic on the face of it ... This device is inherently of no use to us. We do not recommend its purchase.'
WESTERN UNION COMMITTEE, 1876

'There are conditions in America which necessitate the use of such instruments [telephones] more than here. Here we have a superabundance of messengers, errand boys and things of that kind ...'
SIR WILLIAM PREECE, CHIEF ENGINEER, BRITISH POST OFFICE, 1878

'I think there is a world market for maybe five computers.'
THOMAS WATSON, PRESIDENT OF IBM, 1943

'Computers in the future may ... perhaps only weigh 1.5 tons.'
POPULAR MECHANICS, 1949

'Television won't be able to hold on to any market it captures after the first six months. People will soon get tired of staring at a plywood box every night.'
DARYL ZANUCK, CO-FOUNDER, 20TH CENTURY FOX, 1964

'There is no reason why anyone would want a computer in their home.'
KEN OLSEN, FOUNDER OF DIGITAL EQUIPMENT CORP., 1977

'I predict the internet will soon go spectacularly supernova
and in 1996 catastrophically collapse.'
ROBERT METCALFE, INVENTOR OF ETHERNET, 1994

'Amazon is a very interesting retail concept, but IBM is already generating more revenue and certainly more profit than all of the top internet companies combined.'
LOU GERSTNER, IBM CEO, 1999

'Google's not a real company. It's a house of cards.'
STEVE BALMER, MICROSOFT CEO, 2001

'Even with the Mac, Apple attracted a lot of attention at first, but they remained a niche manufacturer. That will be their role in mobile phones as well.'
ANSSI VANJOKI, NOKIA CHIEF STRATEGY OFFICER, 2007

ISBN: 978-1-928257-73-8
e-ISBN: 978-1-928257-74-5

First edition, first impression 2020
Published by Bookstorm (Pty) Ltd
PO Box 4532
Northcliff 2115
Johannesburg
South Africa

www.bookstorm.co.za

Proofread by Sean Fraser
Cover design by mr design
Book design and typesetting by Triple M Design
Index by Steve Anderson

Alan Knott-Craig can be contacted at alan@herotel.com,
or visited at 156 Dorp Street, Stellenbosch, South Africa
www.bigalmanack.com

John Tysoe can be contacted at john@themobileworld.com

Contents

Introduction

'Mr Watson, come here, I want to see you.'

I t's been almost a century and a half since Scottish-born scientist Alexander Graham Bell said these immortal words on the first-ever phone call, to his assistant in the next room.

Between then – 10 March 1876 – and now, the world has changed beyond recognition. And telecommunications, which has played a fundamental role in this change, has itself evolved into an industry that not so long ago was the sole preserve of science fiction.

When the world's first modern mobile telephone network was launched in 1979, there were just over 300 million telephones. Today, there are more than eight billion, most of which are mobile. Most people in most countries are now able to contact each other in a matter of seconds. Soon, we'll all be connected, not just to each other, but also to complex computer networks that provide us with instant information, and also observe and record our actions. There's no other phenomenon that touches so many of us all so directly, each and every day of our lives.

This book is an attempt to describe how this transformation came about. It considers the technologies that underpin telecommunications – micro-circuits, fibre-optics and satellites – and touches on the financial aspects of the industry – the privatisations, mergers and takeovers that have helped shape the $2-trillion telecom market. But for the most part it's a story about us and our need to communicate.

CHAPTER 1

In the beginning

For almost all history, mankind's ability to communicate was limited in both range and content. Prehistoric tribes used fire or smoke signals as warnings, and drums and horns were used in battle, but whatever the medium, it was restricted by the range of human senses: only the largest fire creates a glow that can be seen over the curve of the Earth, while even the loudest drums can't be heard more than a few miles away.

This is not to say that early man lacked ingenuity. On the contrary. Cyrus the Great of Persia established a functioning postal system in the 6th century BCE to facilitate communications across his vast empire. And within a hundred years of Cyrus, other Persians and some Assyrians were using pigeons to carry messages. Crude cryptography appears shortly after, with the Greek creation of the hydraulic telegraph.[1] In China during the Han dynasty (200 BCE to 200 CE) a complex system of flag signals was

---

1 The hydraulic telegraph, a semaphore system invented in about 350 BCE during the First Punic War to send messages between Sicily and Carthage (about 250 miles or 400 kilometres apart, across the Mediterranean Sea), involved a water-filled vessel containing a rod inscribed with identical messages (such as 'horsemen entering the country' or 'ships'). The two communicating parties, which would have an identical set of rods and vessels, would be in sight of each other, usually on a hill. When one wanted to send a message to the other, he would raise a torch, and wait for the other party to raise his torch to confirm he was ready. When the sender lowered his torch, both sides would simultaneously pull the plug from the bottom of the vessel. As the water drained, different messages on the rod would be revealed, and when the intended message was revealed, the sender would again raise his torch, signalling that the receiver should re-plug the vessel and read the message on the rod.

developed, sending messages along the Great Wall and beyond.

For the best part of the next two thousand years the situation stayed much the same, with semaphore and signalling not advancing to any great degree and suffering many of the same limitations. The development of telescopes extended the range of these networks, but that was of little use on a dark or rainy day and no use at all at night.

More fundamentally, the content of the messages sent over these networks was inflexible, as it had to be predetermined and agreed by both sender and receiver. This limitation was slightly ameliorated by the use of coloured flags in naval signalling, but that required a greater level of sophistication on the parts of the operators.

This all changed in the late 18th century thanks to the efforts of various European scientists and inventors. In the 1790s French inventor Claude Chappe came up with the 'tachygraph' (meaning 'fast writer'), an optical semaphore system that transmitted visual signals over a network of physical high points.[2] In 1795 Spanish scientist Francisco Salva Campillo produced a device that transmitted electrical signals representing individual letters over a network of cables.

In 1816 Englishman Francis Ronalds created the first working telegraph over a substantial distance, laying an eight-mile (13-kilometre) length of iron wire between wooden frames and sending pulses down it using electrostatic generators. And in 1832 Russian aristocrat Pavel Lvovitch Schilling created a machine that used a single needle and a system of codes to generate individual characters.

The achievement of producing the world's first truly commercial telegraph network fell to two Englishmen, inventor William Fothergill Cooke and academic Charles Wheatstone, in 1837. Their telegraph used a combination of five needles, each with its own wire, which could point to some

---

2 This invention was predated by several ground-breaking theories, one of which was the suggestion more than 40 years earlier by a Scotsman (known only by the initials 'CM', but probably Charles Marshall) that electrical currents might be passed down a complex of insulated wires to spell out messages; and the work of English polymath Robert Hooke who in 1684 published a theory of telegraphy that included the observation that sound could be made to travel down tightly stretched wires.

20 separate characters arranged in a grid pattern on a board.[3]

In 1839 the Great Western Railway commissioned a Wheatstone and Cooke telegraph line running from Paddington Station in London out to West Drayton, a small town some 12 miles (20 kilometres) to the west. This line is famous for having carried the first telegraph used to apprehend a criminal. In 1845 a certain John Tawell poisoned his mistress in her home outside Slough, before boarding a train to London. Tracked by a member of the local constabulary to the station, his destination and appearance were sent ahead to Paddington, where an arrest was made. Large crowds attended the subsequent hanging and the notoriety of the affair helped push the new technology firmly into the public consciousness.[4]

At around the same time, across the Atlantic Ocean in the United States, portrait painter Samuel Morse was spurred into turning his attention to electrical communications after his wife died in 1825. He'd missed both her death and the funeral because at the time he was away from home completing a commission and the news took too long to reach him. Eventually, in 1837, he patented his invention and made the telegraph his own.

The years that followed were characterised by enhancements to and developments of the basic telegraphic concept and its means of delivery. The first public telegraph line was opened in 1845 between London and Gosport in Hampshire (an important base for Britain's Royal Navy at the time) and the first transmission over this network was Queen Victoria's speech on the opening of parliament. In 1850 the first subsea telegraph was transmitted, from Dover to France, and by 1858, the Atlantic had been crossed, with a cable stretching from Ireland to Newfoundland.[5]

By the middle of the 19th century, telegraph networks were being

---

3 Their design was influenced by the work of two German scientists, Wilhelm Weber and Carl Gauss, who'd created an electric telegraph system in 1833.

4 A similar arrest was made 65 years later of an even more notorious criminal, using what was then cutting-edge technology. Having killed his wife and dismembered her body, Dr Hawley Harvey Crippen, an American homeopath living in London, booked passage on the SS *Montrose* bound for Quebec. His attempts to disguise himself didn't fool the captain, who transmitted the following message by means of Mr Marconi's new wireless telegraph back to the headquarters of the White Star Line in London: 'Have strong suspicions that Crippen—London cellar murderer and accomplice—are amongst saloon passengers. Moustache taken off. Growing beard ...' Police were waiting for him in Quebec and, like Tawell's, Crippen's life came to an end at the end of a rope.

5 It failed after just a month in operation and was not successfully replaced until 1866.

established in most of the world's industrialised countries. In France (and elsewhere in continental Europe) the business of telegraphy had begun as a private enterprise but very quickly fell under the control of the state, nationalised in 1851 by King Louis Philippe and merged into a single entity.

In Britain, by contrast, multiple operators, most privately owned, set up shop. The first, the Electric Telegraph Company (ETC), was established in 1845 by William Fothergill Cooke and his new partner, John Lewis Ricardo. The ETC used the technology patented by Cooke and Wheatstone and grew rapidly, assisted by the expansion of the rail networks, many of which now used telegraphy to facilitate signalling.

In 1850 a second operator, the British Telegraph Company, was launched, using a telegraph system developed by civil engineer Edward Highton. In quick succession, several more newcomers, some with specific regional focus, entered the market, including the English and Irish Magnetic Telegraph Company, the London District Telegraph Company, the United Kingdom Electric Telegraph Company and the Universal Private Telegraph Company.

At the same time, a few special-purpose companies were being formed to provide communications over particular international routes. The Submarine Telegraph Company was formed in 1851 to operate an exclusive concession given it by the French government. A year later, ETC created a subsidiary called the International Telegraph Company to address a similar opportunity in Holland.

By the end of the decade, this trickle had become a raging torrent, thanks in large part to an occurrence that had taken place half a world away from Britain.

News of the 1857 'Indian Mutiny', or First War of Independence, which was started in Meerut by Indian troops in the service of the British East India Company and spread to Delhi, Agra, Kanpur and Lucknow, didn't reach London until a full 30 days after its inception. Although there were telegraph systems in use in India at the time, they were essentially domestic and, once at the coast, messages had to be carried by boat to the next telegraph point, or indeed the whole way. Moreover, the Indians had cut most of the lines at the start of the rebellion.

The British government determined that it should never be caught on the wrong foot again and encouraged entrepreneurs to create companies to provide end-to-end telegraph links. In quick succession numerous international cable companies sprang into being, including the British Indian Submarine Telegraph Company, the Eastern Extension Australasia and China Telegraph Company, the Marseilles, Algiers and Malta Telegraph Company, the Falmouth, Gibraltar and Malta Telegraph Company, and the Eastern Telegraph Company (later to become part of Cable & Wireless).

In the mid-1860s, postal-services reformer Frank Ives Scudamore began campaigning for the nationalisation of the industry, suggesting that the privately held telegraph system in the UK compared unfavourably to the state-controlled monopolies on the other side of the English Channel. Despite protestations from the companies, various Acts were passed and the entire inland telegraph system came under control of the Post Office in 1870. It would take 115 years for this decision to be reversed.

The industry structure in the US was similar to that in Britain. Privately owned telegraph companies, most with limited regional ambitions, proliferated, though none was interconnected. The New York and Mississippi Valley Printing Company, formed in 1851, however, aimed to create a national telegraph network using Samuel Morse's technology. By 1856 it had acquired 11 other networks and its reach extended to St Joseph, Missouri, a distance of over 1,100 kilometres. At this point it decided to change its name to something less regional and more memorable – and became the Western Union Telegraph Company. It dropped the word 'telegraph' a few years later.

When Civil War broke out in 1861, Western Union offered to build the government a chain of interlinked telegraph stations. Almost everyone thought the company was reckless, if not entirely deranged. Abraham Lincoln told Western Union's president Hiram Siley, 'I think it's a wild scheme. It will be next to impossible to get your poles and materials distributed down on the plains, and as fast as you complete the line, the Indians will cut it down.'[6]

---

6 United States History, u-s-history.com/, accessed 3 September 2019.

The task of expanding the network westward from St Joseph began on 4 July 1861 and the line reached Salt Lake City, 1,500 miles (2,400 kilometres) away, on 24 October – Siley's men were covering a distance of over 13 miles (20 kilometres) a day, every day.

In October 1861 a young self-taught German electrician called Philipp Reis demonstrated a device he'd constructed to the Scientific Society of Frankfurt-am-Main. It consisted of a diaphragm that captured sound and converted it into electrical impulses, which it then transmitted over electrical wires to a device that reversed the process and reproduced the original captured sound. Reis called his invention a telephon.

On 14 February 1876, a young Scottish engineer, Alexander Graham Bell, filed a patent in his adopted country, the US, for an apparatus capable of transmitting speech. Mere hours later a similar application was made by Elisha Gray, a superintendent employed by Western Union. Bell was granted US Patent No 174,465 on 7 March and went on to achieve universal, lasting fame. Gray sank into relative obscurity, though he is now considered to be the father of the modern music synthesiser, and was granted over 70 patents for other inventions.[7]

Bell spent much of the rest of 1876 trying to improve his invention and make it a commercial proposition. This proved far harder than he'd expected. His financial backers, Thomas Sanders and Gardiner Hubbard, were beginning to despair and offered their rights in the patent to Western Union for $100,000 (roughly $2,350,000 in current value).[8]

In what could be the most extreme case of technological myopia ever, Western Union declined. The committee set up to consider the offer called the device 'hardly more than a toy' and declared it was 'of no use'. It concluded, 'We do not see that this device will be ever capable of sending

---

7 Gray went on to co-found Gray and Barton, manufacturers of telegraph equipment, which later become the Western Electric Manufacturing Company and eventually just Western Electric. Many years on, it became Lucent and then, after a merger with the French company Alcatel, it became part of Alcatel Lucent. Today, the business Elisha Gray helped found is a subsidiary of the Finnish company Nokia Oy.

8 All $ amounts are US dollars unless otherwise stated.

recognizable speech over a distance of several miles. Messer Hubbard and Bell want to install one of their "telephone devices" in every city. The idea is idiotic on the face of it. Furthermore, why would any person want to use this ungainly and impractical device when he can send a messenger to the telegraph office and have a clear written message sent to any large city in the United States?'[9]

Bell somehow managed to persuade his backers otherwise, and incorporated the Bell Telephone Company in 1877 with Hubbard as president and Sanders as treasurer. Thomas Watson, the man Bell had summoned during that famous first call, took the post of 'general superintendent', while Bell himself had the title 'electrician'. The Bell Telephone Company began, at last, to gain some traction, with licences granted to independent operators to establish networks in New York, Boston and Chicago.

In 1878 a second company, the New England Telephone Company, was licensed to operate in the six states that comprise New England.

It didn't take Western Union long to realise that it had made a huge blunder. When one of its subsidiaries abandoned the use of telegraphy in favour of Bell's invention, it recognised that the telephone represented an existential threat, so it quickly hired the brightest people it could lay its hands on to create a company that could start up in competition.

In December 1877 the American Speaking Telephone Company, a business whose sole purpose was to compete with Bell, was established, and Elisha Gray was brought in, as was a young inventor called Thomas A Edison. Edison brought with him a design for a transmitter that was far better than the one used by Bell and this, together with Western Union's existing network of 250,000 telegraph cables, could have proved enough to kill the new business before it had really begun.

Bell's company took the only real option available to it and filed a suit for patent infringement in early 1878. The case would run for the best part of two years, before Bell eventually emerged victorious in November 1879. The settlement saw Western Union relinquish all of its telephone patents and hand over the 56,000 phone lines it had connected, expanding Bell's

---

9 History of AT&T, Cybertelecom.

installed base to some 133,000 phones. In return, Bell agreed not to enter the telegraph business and to pay Western Union a 20% share of Bell's income from the rental of telephone handsets for the remaining 17 years its patents had to run.

In 1879 a new general manager, 33-year-old Theodore N Vail, took up the reins at the Bell company. Vail's vision was at variance with the prevailing mode of the day: his primary objective was to provide a good service rather than maximise profits, as he believed that good service would eventually lead to good profits. A straight shooter, he also believed in long-term planning and strategic thinking. Among his first moves, in 1880, was to rename the company the *American* Bell Telephone Company.

In 1881, Vail negotiated the acquisition of Western Electric, a business Elisha Gray had helped establish and which was becoming an increasingly important force in the industry. Just five years after its inception, Bell's original entity had become a vertically integrated, full-service company, capable of addressing all the needs of the new industry and its customers. It changed its name once more, calling itself the American Telephone and Telegraph Company, or AT&T.

The business of telephony was then – and still is now – very capital intensive. The equipment is expensive and a considerable amount of it has to be in place before the first revenue can be earned. Moreover, networks have to be built to handle the level of traffic at the one moment of the day when demand is at its highest: for the rest of the day, the equipment is effectively under-utilised. The young AT&T just didn't have the resources to fund its own expansion.

Vail sought to address this by raising additional capital in 1880, but it wasn't sufficient. The expansion of the network continued to consume large amounts of cash. Vail concluded that it made more sense to rent what is now called customer premises equipment (CPE) to subscribers, rather than sell it to them. The rental arrangement ensured a long-term revenue stream and a long-term relationship with the customer, both of which he saw as immensely important, but it also placed demands on cash flow, which, at this stage in its development, the young company wasn't easily able to address.

From the very beginning, the Bell Company had offered licences to independent operators to reduce the need for capital. These licences generally ran for 5–10 years, after which Bell had the right to acquire the operator. Vail changed this approach in 1881, offering permanent licences to independents in exchange for equity stakes of anything between 30% and 50%. The independents were permitted to connect individual exchanges to each other with long-distance lines but weren't allowed to connect directly with other independents' networks. This approach allowed it to deploy most of its capital where it was most needed, in the expansion of the long-distance network.

In the late 1880s the company's investors, and especially its early backers in Boston, wanted better returns, quicker. The long-term aim of building value through a high level of service was questioned and, as a result, in 1877 the excellent Vail left the business.

The years that followed were difficult for AT&T, with independent networks springing up in parts of the country they didn't serve (notably the West and Midwest), construction costs rising faster than anticipated, and quality of service declining along with the company's reputation.

None of this prevented a sharp rise in demand for telephones, however, and in the years leading up to the end of the century, AT&T continued to expand its operations. By 1900 it covered much of the urban population of the US. It had 600,000 telephone lines in service, accounting for around three-quarters of the national total. This equated to a penetration rate of over 1% of the population – by far the highest level in the world. Most of these lines were still urban, but rural uptake was rising fast, and by 1907 it had increased to 1.4 million lines, nearly half of the company's total base of 3.12 million subscribers.

The telephone wasn't an exclusively American preserve. Bell's device sprang up in South America, Europe, Asia and Africa, all within ten years of its invention – an incredibly impressive achievement considering that there was no advertising or other media to circulate news of the invention, and it spread itself around the world on the basis of its own merits alone.

The Brazilian emperor Pedro ll apparently saw Bell's invention at the

Centennial Exhibition in Philadelphia in 1876, tried it, and famously exclaimed 'Dios! It works!' By the end of that year, some months before the incorporation of the Bell Telephone Company, he'd installed one in his palace in Rio. His early interest helped establish a huge telephone industry in the country – by 1940 there were over 800 separate telephone operators scattered across Brazil's vast expanse.

In 1877, one year after Bell had acquired his patent, telephones were making their first appearance in several European countries, including France, Germany, Portugal, Spain, Sweden, Switzerland and the UK.

In France, Téléphone Bell was established by Cornelius Roosevelt, the paternal grandfather of the future US president Theodore, in 1877. The following year Roosevelt met Frederic Allen Gower, an American entrepreneur and engineer who'd made some modifications to Bell's original instrument and patented the changes as the Gower-Bell telephone. Gower and Roosevelt agree to a merger. Two years later, in 1880, the Compagnie du Téléphone Gower made its first acquisition, buying the smaller Société Française de Correspondance Téléphonique de Soulerin (named after its founder, a French engineer called Leon Soulerin), which had been granted a licence to operate in Paris the previous year.

In 1878 Thomas Edison created the Société du Téléphone Edison with licences to operate telephone systems in Paris, Lyon, Bordeaux, Marseilles, Nantes and Lille. Within a year it had merged with Compagnie des Téléphones to form the Société Générale des Téléphones (SGT). SGT was nationalised in 1889 and would remain a branch of government for 101 years and wholly owned by the state for a further eight, before it emerged as France Telecom, ready for its partial privatisation in 1997.

On the other side of the English Channel, meanwhile, the first British telephone company came into existence on 14 June 1878. The Telephone Company Ltd (Bell's Patents) initially concentrated its efforts on the sale of telephone instruments and the installation of private lines, but in August 1879 it opened what is believed to be Britain's first public telephone exchange (basically, a room in which human operators interconnected telephone lines to establish telephone calls between subscribers) at 36 Coleman Street, in the financial district, a few minutes' walk from the Bank of England

and the London Stock Exchange. It followed this later in the year with a further two exchanges in London and seven regional offices, in Glasgow, Manchester, Liverpool, Sheffield, Edinburgh, Birmingham and Bristol. On 15 January 1880 it issued the first known British telephone directory, listing 250 subscribers connected to the three London exchanges.

In August 1879 the Edison Telephone Company of London Limited was floated, and Edison promptly sued The Telephone Company, claiming the same infringement of patent rights that Bell had brought against Western Union in the US. Following the settlement of Bell's litigation in the US, their British counterparts agreed to settle their differences too, and on 13 May 1880 the United Telephone Company was formed through the merger of The Telephone Company Limited (Bell's Patents) and the Edison Telephone Company of London.

However, later that year a case was brought by the Post Office regarding the definition of 'telegraph' under the Telegraph Acts. Despite the chief engineer of the Post Office having declared the new American contraption not suitable for use in the UK due to the latter country's 'superabundance of messengers, errand boys and things of that kind' to fetch and carry communications,[10] the judge ruled that a telephone was a telegraph and a telephone conversation a telegram.

One implication of the ruling was that, technically, the United Telephone Company – and all of its smaller competitors – ought to be merged and come under the control of the state, but the government chose not to push the point and came to a compromise. The business, together with all others in the same field, was to operate under a licence awarded by the Post Office for a period of 31 years, at which point the Post Office had an option to acquire the assets and businesses of the telephone companies. The United Telephone Company was issued with its licence on 1 January 1881.

A year later, however, the postmaster general, Henry Fawcett, had a change of heart, judging that it wasn't in the public interest for the telephone service to be a monopoly. This reversed the original 1869 decision, the one that had given the Post Office exclusive rights to provide telegraph

---

10 1879 Report of the Select Committee on Lighting by Electricity.

services. Fawcett announced that from that point on, licences to operate telephone systems would be awarded to 'all responsible persons' who applied for them, even where a Post Office system was already established. Two years later, he went even further, removing all restrictions on telephone operators' service areas. Previously, these had been limited to an area within five miles (eight kilometres) of the exchange. This decision allowed companies to begin creating multiple exchange areas, connected by trunk (long-distance) wires and, potentially, a unified national system. This, of course, was exactly what Bell was trying to do in America.

At the same time, Fawcett gave telephone companies the right to establish public telephone stations in both private and public places. New 'call offices' were soon to be found in 'silence cabinets' in shops, railway stations and many other public locations, an inspired move that demonstrated to a sometimes sceptical public that telephones actually worked. At a stroke, Fawcett had opened up a vast additional market for the service.

In Sweden – a country that has played an important role in the development of the global telecommunications industry – the first phone line was laid in in 1877 by Kongliga Elektriska Telegraf Verket (the Royal Electric Telegraph Works). This state-owned body had been created 24 years earlier, when the first telegraph line was installed. Swedish legislation suggested that the industry should be state owned and a monopoly, but neither of these rules was applied with any great vigour.

In 1883, three years after the first exchange was installed in Stockholm, Swedish engineer Henrik Tore Cedergren formed a company called Stockholms Allmanna Telefonaktiebolag (the Stockholm Public Telephone Company or SAT). This was a private enterprise, despite the state's ownership of the telegraph network and its belief that telephony ought to be similarly treated. SAT was to remain in private hands until 1953, when the state took the decision to nationalise the entire industry through the creation of Televerket, which today operates as the Telia Company.

Cedergren's ambition was to put a phone in every household in Sweden, and he began collaborating with the new company Lars Magnus Ericsson had formed. Eventually, in 1918, the two businesses merged to form Allmänna Telefonaktiebolaget LM Ericsson (the General Telephone

Company LM Ericsson.)

Ericsson had started out as an instrument maker, working for Televerket, and had then gone into business for himself, repairing telegraph equipment. In 1878 he was asked by a local importer of Bell's telephone devices to make adjustments to improve their performance, and Ericsson was soon manufacturing phones based on the Bell design with a few of his own modifications. He was able to do this without the payment of royalties or the risk of patent infringement as Bell hadn't thought to take out a patent in Scandinavia.

In the German Empire, all telephone operations in the country had been declared to be under the control of the postal authority in 1877, while in Russia, the government issued a decree in 1881 that the development of an urban telephone network should be permitted, with the state becoming increasingly involved from 1884 onwards.

Farther afield, the first telephone service was available in Japan as early as 1877 – but only to employees of the government, public bodies, the police and a handful of selected private businesses. The public had to wait until 1890, when lines were laid between Tokyo and Yokohama.

The first telegraph networks were built in China in 1877 to provide better communications for the military in the event of possible future hostilities with the Japanese. A submarine cable linked the Chinese mainland to the island of Formosa (now Taiwan, or the Republic of China) to which the Japanese had laid claim. When the Chinese lost the first Sino-Japanese War in 1895 it had to cede control of Formosa to Japan, and the excellent communication system that had been built up ahead of foreign aggression now, ironically, fell into the hands of the aggressor.

In several countries around the world the first telephones were introduced by the British, in part to facilitate colonial administration. In Malaya (today's Malaysia) the first phones were introduced as early as 1880, and by 1891 had appeared in the capital Kuala Lumpur. In India the British established the Bombay Telephone Company in 1882 to provide local service and connectivity with the Empire. It still exists today, as Mahanagar Telephone Nigam Ltd (MTNL).

According to Telkom South Africa, the first dozen telephones arrived in

South Africa in 1878. They were a gift from the Bell Telephone Company, sent to the new commander of the British forces, Sir Garnet Wolseley, to assist in the Zulu War. The Post Office in the Cape Colony realised the importance of the new invention almost immediately, and by 1880 a commercial service was in place.

From the very beginnings of the industry, engineers and inventors strove to improve Bell's invention and enhance its utility. In 1878 Edison patented a carbon transmitter, a marked improvement on Bell's original magnetic-current design. In the same year someone working for the US Coastguard utilised unpatented designs by the British inventor David Hughes to create something similar – which was promptly adopted by the Bell Company and used in its telephone instruments for the next 20 years. (Its eventual replacement was also invented in the same year– and patented – by an English clergyman, the Reverend Henry Hunnings of Bolton Percy, Yorkshire.)

Hughes followed up his 1878 invention with another key development when he transmitted 'aerial electric waves' from his home in London to a receiver some 450 metres away. This was the first wireless transmission ever, exactly 100 years before the Nippon Telegraph and Telephone (NTT) cellular service was launched in Tokyo. It was also 14 years before Nikolai Tesla's first efforts (which some claim to be the first wireless transmission) and some 17 years before Guglielmo Marconi's successes in the same field.

Telephone instruments became more reliable, with both transmitter and receiver greatly improved. Bell's colleague Thomas Watson had invented a new ringing mechanism that allowed the instrument to alert subscribers to an incoming call, while Lars Ericsson had managed to combine the two units into a single handset in 1892.

The invention of the first telephone exchange was a momentous development – before this, all lines were connected to only one other line, and if you wanted to talk to more than one person, you'd need more than one phone.

There's some dispute as to exactly who should take the credit for this.

Tivadar Puskás, a Hungarian who was working for Thomas Edison, is thought by many to be the best candidate for the honour, having opened a telephone exchange in Paris in 1879. Others assert that it was the German postmaster Heinrich von Stephan who opened the first European exchange in Germany in 1877. There's less controversy about the world's first commercial (revenue-generating) exchange, which was opened in 1888, having been designed by an American, George W Coy. Charles Glidden, an early automotive pioneer, is also credited with building an exchange, in Lowell, Massachusetts, in the same year, but his more lasting contribution to the industry was that he was the first to realise female voices worked better over the new invention than male. He duly staffed his exchange exclusively with women.

In 1883 a Scottish engineer at the National Telephone Company's Glasgow office patented an automatic telephone exchange, which allowed a subscriber at a branch exchange to be connected to the central exchange without manual intervention at the branch level. This was a big step towards one of the biggest single steps in the history of the industry – the invention of a fully automatic exchange.

Kansas City in the latter years of the 19th century was a fairly dangerous place to live and many in the town failed to do that for very long. The high mortality rate among the 132,000 citizens ought to have been sufficient to support at least two top-class funeral parlours. So why was the one owned by Almon Strowger not flourishing? Close scrutiny of the matter revealed the reason: Mr Strowger's main competitor was married to the woman who operated the local telephone exchange, and she was diverting all calls meant for Almon to her other half.

Strowger was determined to eliminate this unhelpful intervention. He began experimenting with brass collar studs, metal pins and the like in 1888 – and by 1891 he had created a machine that would bypass his competitor's interfering wife. He was awarded a patent for the device (also known as a step-by-step or SXS switch) in the same year.

Together with a few relatives he then founded the Strowger Automatic Telephone Exchange Company. He and his partners began marketing the new invention to the various telephone companies that were now scattered across the country with the accurate if somewhat wordy slogan the 'girl-less, cuss-less, out-of-order-less, wait-less telephone'.[11]

Although the Bell System was reluctant to embrace the Strowger exchange, preferring to use manual exchanges of its own design, many of the country's small independent operators took to the device with alacrity. The first Strowger switch was deployed in the same year as its invention, in the small town of La Porte, Indiana. It was soon to be seen in exchanges across the whole of the US, albeit usually in places some way off the beaten track.

The credibility of the company (by now known somewhat ambiguously as the Automatic Electric Company) was greatly enhanced in 1912, when the system was selected by the British Post Office as the principal mechanism to serve small or medium-sized towns, in preference to other exchanges offered by such well-established manufacturers as Western Electric, LM Ericsson, Siemens and Lorimer. (This last has now fallen into obscurity but for a while it was a contender. A North American manufacturer, it would eventually be acquired by Bell's Western Electric subsidiary.)

For its part, BT continued using Strowger exchanges in its network right up to the mid-1990s, when the last of several thousand local exchanges was phased out and replaced with a more modern digital system.

In 1896 Italian electrical engineer Guglielmo Marconi visited the British Post Office, wanting to talk about a new system of 'telegraphy without wires' that he'd invented. Marconi had already approached the relevant bodies in his native Italy but they hadn't been interested.

The British engineer-in-chief was, however, and a demonstration was arranged. Marconi rigged up a transmitter on the roof of the London Central Telephone Exchange and successfully beamed a signal to a site some 300 metres away. The Post Office subsequently allowed Marconi to undertake large-scale experiments, on Salisbury Plain and elsewhere, and provided the entrepreneur with finance. A patent was granted for the invention the following year.

---

11 This last alludes to the lengthy process needed to establish long-distance calls. With the older manual systems an operator at one exchange would have to contact her counterpart at a more distant exchange to make the connection between exchanges, the second person, in turn, might have to call a third, and so on, until all the necessary connections had been established. This typically took about a quarter of an hour.

By 1899 Marconi's wireless systems were powerful enough to cross the English Channel, a distance of some 20 miles (32 kilometres). Two years later the engineering had advanced sufficiently to allow a signal to be transmitted across the North Atlantic, from Cornwall to Newfoundland.

By the beginning of the 1900s the concept of the telephone had firmly established itself, at least in the developed world. The industry looked set to continue to grow, with many businesses using it in preference to the older technology of telegraphy.

CHAPTER 2

New technologies for a new century

In the first decade of the 20th century, AT&T was losing the battle against its independent rivals, which now owned over half the exchange lines in the USA and were offering rates that were well below those AT&T would have preferred to charge. The company's balance sheet was showing increasing levels of debt, and this more than tripled from $60 million to $200 million between 1902 and 1906.

The financier JP Morgan was very quick to spot an opportunity, and in 1903 he began to acquire AT&T's publicly traded bonds. By 1907 he and his associates in New York and London had wrested control of the business from the Bostonians with whom Theodore Vail had fallen out two decades earlier.

The new controlling shareholders agreed that the culture of customer service Vail had fostered had to be re-established, and that the best way to achieve this was to bring back the man himself. In 1907, therefore, Vail began his second term as president of AT&T. The benefits of the changes he wrought weren't immediate but in the long run they were profound and very beneficial to customer and shareholder alike. Within a decade the debt had been brought under control, service quality had improved beyond all recognition, and a secure path to the future was being established through an emphasis on technological research, an effort that resulted in the creation of the world-renowned industrial research and development outfit Bell Telephone Laboratories in 1925.

Although Vail did much that's laudable, he remained convinced that the business of telephony ought to be a monopoly. Together with Morgan, he did everything in his power to make AT&T a monopoly along European lines. (Britain's postmaster general had deemed competition in the phone industry to be 'wasteful' as early as 1901 and Vail, it seems, agreed with him.) In 1910 he took over the role of president of Western Union, after AT&T had acquired a 30% interest in its former rival. The move created what would now be called 'synergies', as for the first time telegrams could be sent over telephone lines, creating a second revenue stream from the same set of assets.

At the same time, AT&T continued to acquire as many independents as it could. Morgan was quite prepared to use the financial muscle at his disposal to force struggling companies to surrender their independence to AT&T, making the process of expansion relatively easy. This kind of behaviour tends to generate a backlash sooner or later, and so it did here. Consumers and government alike began to wonder whether AT&T was breaching certain aspects of the country's 'antitrust laws', a collection of federal and state government laws that regulated corporations in order to promote competition for the benefit of consumers. The question was raised before the First World War; the answer was 70 years coming, but when it did, it was 'yes'.

One of the complaints levelled against AT&T was that it didn't allow competitive local-exchange operators to access its long-distance network. If they wished to phone anyone outside the immediate community, subscribers to an independent company would have to double up and take Bell service as well. The numbers at the leading New York state independent Rochester Telephone are typical: by 1907 it had a total of 10,000 subscribers, but Bell, where charges were significantly higher, had over 14,000 in that same region.

This issue irked consumers and the state. Vail thought that a gesture was needed to help the antitrust pressures disappear, so in 1913 he opened up access to the long-distance network to AT&T's competitors, and at the same time he disposed of the company's stake in Western Union. AT&T's attempt to create a national monopoly had failed – but looking at the position in Europe, that was probably no bad thing. The state-controlled monopolies on the other side of the Atlantic served far fewer customers.

Since its inception, AT&T had made every effort to stimulate demand for the telephone. The motivation is obvious: Alexander Graham Bell had invented it, and he and his company had a huge vested interest in promoting it – a far greater interest than the CEO of a public utility might be expected to show, however dedicated they were. A comparison with the position in Britain makes this point clear. At the time that the Post Office took control of the National Telephone Company and the other UK independent operators (in 1911) it had a total base of about 77,000 lines, accumulated over the years since it had first entered the market in 1878. National by itself had over seven times as many subscribers (561,000) connected over the same 33-year period. This might suggest that businesses should be run by private enterprise rather than government – and as we shall see, evidence from other markets points to the same conclusion.

At the outbreak of the First World War, the total number of lines run by the combined operations had risen slightly, to around 730,000 lines. This is equivalent to a penetration rate of just 1.7% – way below the level in the US, which had nearly 10 million lines serving its 100-million population. Apologists for the concept of state ownership (and there are still a few) suggest that the poor performance in the UK was due to a lack of investment by the private companies, combined with indifferent service and high tariffs, rather than the businesses' ownership structures, and there may be some truth in this. In the years leading up to the handover to the Post Office, there was very little incentive for the managers of the independents to run their businesses with a view to the long term, as they were going to lose ownership of the business in a couple of years. That such a situation was allowed to come about can't be laid at the door of private enterprise.

This 'explanation' also doesn't account for variations between state-owned enterprises in different countries. Telephone penetrations varied hugely between the world's leading economies in 1914. The US with its privately owned networks was well ahead at around 10%, with Canada (also a private-enterprise market) in second place at just over 6%. New Zealand (state owned) came in third at 4.6%, while of the others, only Sweden (theoretically state controlled) edged above 4%. Of the others, five were state-owned monopolies, the exception being Japan where the population of 54 million was served by less than a quarter of a million lines.

A possible explanation for these differences is that in those early days, telephones were used by those who had the greatest need for them. The British, as noted, had servants to deal with the tiresome matters of the day, but elsewhere, the farther you were from your nearest neighbour, the more you needed help to communicate with them.

Technology continued to improve. New components and better, copper-enhanced sound quality reduced the need for signal amplification. Simple forms of error coding increased the resilience of telegraph traffic, while the quality of voice traffic was enhanced by the development of audio filtering and 'companding'. (Companding, or compansion, is a means of improving the quality of a signal, through selective *comp*ression and exp*ansion*). Analogue telephone lines of this kind provided low cost connectivity for the best part of a century before being replaced by more efficient digital technology.

New milestones were passed at an accelerating rate: the first transcontinental call was made in January 1915 when Alexander Graham Bell in New York called his former collaborator Thomas Watson in San Francisco. By October that year, the Atlantic had been crossed by the first wireless call.

Efforts to improve on Strowger's exchanges continued throughout this period. In 1915 Western Electric achieved something of a breakthrough with the design of a 'coordinate selector', which was to form the basis for the first 'crossbar' switching. Crossbar switches are based on a matrix design, allowing any input to be connected to any output. It wasn't much used in the US at the time, but in 1922 Swedish national Gotthief Angarius Betulander, who was employed by Telegraf Verket, adapted the concept to create a new kind of exchange, using electromagnetic relays to perform the switching function. Subsequent enhancements of the design by both Western Electric and LM Ericsson were to follow and, in the years after the Second World War, crossbar exchanges increasingly replaced Strowger designs in many markets around the world.

The structure of the industry and its operating dynamics had changed little over the war years, though consolidation was still very much a feature in the United States. AT&T accounted for around half the phone lines in the country, while there wasn't really a number two at this point, as the industry was fragmented and the next-largest business was a small fraction of the size of the market leader.

Credible data from this period is pitifully scarce – most of these businesses just didn't keep very reliable records – so to put this into perspective we have to look well into the future, as it was at the time. By 1955, four decades after it had begun acquiring other independent telephone businesses, General Telephone (later GTE – General Telephone and Electric) had become the largest independent operator behind AT&T. In that year it became even larger by acquiring the second largest, Theodore Gary & Company. Did AT&T feel the hot breath of competition on its collar? Hardly – the deal increased General Telephone's customer base from 1.9 million telephones to 2.5 million, but the extra 600,000 lines barely made a ripple on the surface of the market; at the same time, AT&T was operating well over 30 million lines, with over two thirds of all homes subscribing to its service.

Three decades after that, in 1983, just before AT&T's breakup, the bottom end of the market was still fragmented, and AT&T still controlled the vast majority of all connections with a total of nearly 90 million phone lines. GTE was still in second place, albeit a long, long way behind, with 15 million. Continental Telephone, Sprint, Southern New England Telephone, Alltel, Rochester Telephone and Cincinnati Bell (no relation) were the only ones to register on the radar. Beneath them there were hundreds of smaller companies, some serving only a single community. That's changed over the last 35 years, but not beyond recognition.

Back in 1918, the sheer power of the beast attracted government attention. The nation's largest operator had 10 million customers, and had put its shoulder to the plough, providing communication services in France for the few months American troops saw action in the Great War while providing sterling service for the top brass back home.

Then the government decided to nationalise it. In 1918 AT&T was made a branch of the US Post Office, the same arrangement the politicians saw in operation in Europe, with the big difference that day-to-day management of the business remained the responsibility of the company. The government had promised that public ownership would bring lower rates, but immediately introduced a connection fee, while also increasing both local and long-distance charges. By August 1919 support for government ownership had entirely collapsed and the brief flirtation with socialism was over.

The First World War saw huge growth in the use of the telephone, both at home and near the front line. The number of phones installed in each of the previously mentioned developed-world markets increased by a minimum of 25%. By 1920 over two thirds of all lines were still in the USA, where the total had risen to over 16.5 million.

The phenomenon wasn't confined to the developed world, however. By the beginning of that decade the phone had also reached such disparate places as the Belgian Congo (now the Democratic Republic of the Congo, DRC), Ceylon (now Sri Lanka), French Guiana, French Indochina (now Vietnam), Iceland, Ireland, Madagascar, Mozambique, Senegal, Somalia, Sudan, Tunisia, Uruguay and the West Indies.

By the end of the decade the number of phones in service around the world had increased to more than 30 million. Albania, Angola, Cameroon, Cape Verde, Cote d'Ivoire, Cyrenaica (now part of Libya), Czechoslovakia, Erythrea (now Eritrea), Kenya, Morocco, Southern Rhodesia (now Zimbabwe), Tanganyika (now Tanzania), Uganda, Upper Volta (now Burkina Faso) and Yugoslavia had all joined the club, making telephony effectively a global phenomenon – but a global phenomenon still dominated by the US, which accounted for over 20 million lines, still some two thirds of the total.

Not all of these networks were based on fixed connections. Wireless telephony was alive and well and was being deployed in some countries where the cost of fixed lines would have been prohibitive. In Chile, for instance, the country's unusual topography precluded the use of fixed lines outside the main metropolitan areas of Lima and Valparaiso.[12]

By the late 1920s average global penetration had reached 1.5%, up from 1% in just a decade. The utility of the telephone system was improving all the time, as more subscribers connected, while the technology in the networks and quality of the subscriber equipment was heading in the same direction.

Then, in late October 1929, the Wall Street crash of 'Black Monday' changed everything. The number of telephones in the country had passed the 20-million mark in 1929, and even through 1930 the growth continued, albeit at a markedly slower rate. Then the growth stopped altogether and the industry began to go into reverse. By 1934, the number of lines in the US had been reduced to just under 16.9 million, a 16.5% drop from the 1930 peak. In just four years, the US lost more telephone lines than there were in the whole of Germany, the world's second-largest market. (It would be another five years before the US got back above 20 million, just in time for another world war.)

AT&T suffered as the largest provider, but the crash almost eliminated some of its smaller rivals. Rochester Telephone, for example, which had had over 100,000 subscribers in 1929, lost all but 11,051 of these.

The impact of the crash was felt in Europe and other developed-world markets, but not to anything like the same degree. Only Germany experienced a decline in telephone ownership, though the losses were modest.[13]

By the end of the 1920s, the telephone systems in most developed-world countries (with the exception of the USA and Japan) were operated by

---

12 In 1928 a new, privately owned company, Vía Trans Radio Chilena Radiotelegraphy Company (VTR) was formed to provide a phone service over fixed wireless links. It was a joint venture, owned by Marconi Wireless Telegraph Company, the Compagnie Generale de Telegraphie sans Fil, Marconi's Belgian subsidiary, Germany's Telefunken and the Radio Corporation of America (RCA), a company that had been the Marconi Wireless Telegraph Company of America, another subsidiary of the British company, but which had been expropriated by the US government during the First World War.

13 A gain of 132,000 lines in 1930 was entirely eliminated the following year, after which 1932 saw the biggest drop of 153,000. This reduced the national total to 2.96 million. These losses were soon offset, however, as Adolf Hitler cranked up the economy, and in the following six years leading up to 1938, nearly 1.2 million lines were added. This took the total above the 4-million mark. Over the same period, the UK more than matched the German achievement, adding 1.25 million lines to give a total of 3.2 million, while the French market also grew, though only by a modest 436,000, which brought its pre-war total to 1.59 million lines.

a state-controlled entity, which often, if not invariably, had links to the country's postal system. In other markets, there was less uniformity. Some operators were controlled by the ruling colonial power, either directly or through an entity like Cable & Wireless or its French counterpart, France Cable et Radio.

Cable & Wireless was formed in April 1929 through the merger of two competing international communications businesses – the cable systems owned by the various British telegraph companies and the General Post Office's international communications interests. These included two cables from the UK to Canada, which the Post Office was obliged to lease to the new company for a period of 25 years, but the mainstay of the business was the 'wireless telegraphy' activities of the erstwhile Marconi Wireless Telegraph and Signal Company, a business that the UK government had first nationalised in 1904. The merger between 'Cable' and 'Wireless' had been arranged by the UK government following the Imperial Telegraph Conference of 1928, when it had become apparent that the cable operators were considering abandoning their international activities in the face of increasing competition from the newer – and cheaper – wireless technology. Believing that the international cables were of strategic national importance, the government concluded that the only way to keep them under British control was to merge them with their competitor.

Cable & Wireless's early development is mirrored in the annals of another multinational telephone business, the American-owned International Telephone and Telegraph (ITT). Few companies in the industry have such a questionable reputation as this business, but be that as it may, it became a major force in the industry, especially in its contribution to the industry's technology. ITT was formed in 1920 by the brothers Sosthenes and Hernand Behn, with the aim of building a global telephone network. The two men had acquired the Puerto Rico Telephone Company and the Cuban-American Telephone and Telegraph Company in 1914, so they had some experience of running telephone businesses. Their first major move on to the global stage came in 1924, when ITT took a controlling stake in Compañía Telefónica Nacional de España (CTNE), a company that had been formed to consolidate several regional operators in Spain. (Today, this company is known as Telefónica.)

The Behn brothers were ambitious and followed this a year later with an even bigger deal. To keep the antitrust hounds at bay, AT&T had agreed that its Western Electric manufacturing subsidiary should divest itself of its international assets. ITT bought them, including the Bell Telephone Manufacturing Company (BTM) of Belgium and British International Western Electric, which was renamed Standard Telephones and Cables (STC). By 1930, ITT had acquired several other European businesses, including two German businesses, Standard Elektrizitätsgesellschaft and C Lorenz AG, a manufacturer of electronic equipment. Through its acquisition of Lorenz, ITT became the indirect owner of a 25% interest in a company called Focke-Wulf Flugzeugbau GmbH, which made aircraft that were to play a significant role in later events.

By the end of the 1930s telephone instruments were generally reliable, transmission systems could reach across oceans, both wirelessly and through cable, and automatic switching was being widely adopted in the developed world.

Operating efficiency was enhanced by the use of 'multiplexing', a technique by which the capacity of a transmission link could be optimised. From the very early days of the telegraph, engineers had noticed that there were long pauses between the pulses sent down the line. (The telegraph wires were not in continuous use and were frequently entirely unused.) They realised that if multiple signals could be woven together, and then unwoven at the other end, the cable could carry much more traffic.

There's a dispute about who should take the credit for this development. An American called Moses G Farmer is known to have successfully demonstrated a duplex telegraph as early as 1853. French sources support the 1870 work of Jean-Maurice-Émile Baudot, who probably made the largest contribution by using synchronised clockwork-powered switches to successfully demonstrate the simultaneous transmission of five separate messages. (In 1877, Baudot was sued by one Louis Victor Mimault, who claimed he'd invented a five-line telegraph system first; Mimault, who seemed to be deranged, was subsequently found guilty of murder and his lawsuit was ultimately unsuccessful.)

As well as multiplexing, transmission technology benefited from research into different areas of the electromagnetic spectrum. In 1931 the world's first microwave link was demonstrated over the 20 miles (32 kilometres) of the English Channel that separates Dover and Calais by ITT's British subsidiary, Standard Telephones and Cables. Microwaves are traditionally defined as radio waves extending from a wavelength of 1 metre (300 MHz) to 1 millimetre (300 GHz). In practice, the term is now usually used to describe point-to-point links at 28 GHz and above. These higher frequencies require a line-of-sight link and can suffer from high propagation loss and signal scatter, but they can be focused into narrow beams, which improves signal strength and reduces interference to other users, which in turn increases network capacity.

Microwave technology was to benefit hugely from developments during the Second World War, especially in the field of radar. The concept dates back to 1865, when the Scottish scientist James Clark Maxwell suggested that electromagnetic energy might be deflected or reflected off an object in the same way that visible light is, a theory proven in the 1880s by German scientist Heinrich Hertz. The science took a huge leap forward in 1940 when John Randall and Harry Boot created a 'cavity magnetron' at the University of Birmingham. This device was capable of producing electronic pulses of sufficient power for effective long-range ground-based radar, and, later, an effective airborne radar system. Today, microwave-based equipment is ubiquitous, not just in radar and in the kitchen, but in devices like mobile phones, wireless local area networks (LANs, which link devices within a building or group of buildings, especially within a radius of less than a kilometre) and Bluetooth connections.

Meanwhile, the multiplexing concept pioneered by Farmer and Baudot had been further developed, and in 1903 WM Milner demonstrated that the technique of signal sampling used in time-division multiplexing could also be applied to telephony.

In 1926 Paul M Rainey of Western Electric made and patented a facsimile machine that converted images into digital signals, using a technique called pulse code modulation (PCM). It's the key digital technology that enabled compact discs to be made and also features in computer disks, digital audio and telephony, and many other modern applications – most of us use equipment based on this technology almost every day of our lives. (Americans Bernard M Oliver and Claude Shannon were granted a patent for their work on PCM technology in 1956.)

In 1937, independently, British engineer Alec Reeves invented a means of using PCM to sample and encode telephone signals. At the time he was working for a branch of ITT in France, but he soon moved to the British government's Telecommunications Research Establishment. This was the main centre for work on radar, radio navigation and other techniques with military applications during the Second World War.

Over the next few years, further digital techniques were developed that would lead to a complete change in the way telephone exchanges worked, while also helping shorten the war by months or even years.

The 'Enigma' machine, an encryption device used in all branches of the German military, looked like a complicated typewriter and that, in part, was because it was made by the famous German typewriter manufacturer C Lorenz AG. Its manufacturers and users thought it so sophisticated it could never be cracked. However, the British managed to capture an Enigma machine along with certain code books, and efforts were made to crack the underlying cryptographic process. The mathematical genius Alan Turing was brought in to Bletchley Park, the then top-secret home of the codebreakers, to perform statistical analyses of the data, and he turned to the Post Office's research branch for help in automating the process. Here, in 1941, Turing encountered Tommy Flowers, the head of the Post Office's switching group. Max Newman, a colleague of Turing's who was in charge of the process of automation, co-opted Flowers.[14]

In the years immediately before the war, Flowers had been working on

---

14 Max Newman became professor of mathematics at Manchester University, where he continued his work with computers. His department set about developing a fully programmable electronic computer and in 1948 the Manchester Baby was born, followed nine months later by the more powerful Manchester Mark 1. By the early 1950s Newman's department was collaborating with the Manchester branch of the Ferranti Company, and in 1951 the two produced the world's first fully programmable, general-purpose, commercial computer, one month ahead of the American UNIVAC 1.

the idea of an all-electronic telephone switch. On his arrival at Bletchley Park, he wasn't impressed by the electromechanical counting machine being used for the job of trying to break the Enigma's code, and suggested he could build a fully electronic equivalent that would be faster and more reliable. Flowers deserves the credit for much of the design of Colossus, the computer that took on the job of deciphering the vast numbers of signals sent out every day by Nazi Germany.

After the war, Flowers resumed work on his electronic digital-switching apparatus. All the switches in use around the world were based on analogue designs where the sound of a voice was converted into an electronic waveform. Flowers wanted to create a digital switch where the voice (or any other kind of signal) would be converted into a digital bitstream (a sequence of bits, or binary sequence), which would improve sound quality and reduce noise, enable encoding of the signal for enhanced security, reduce use of bandwidth during transmission, and enable longer transmission ranges.

Flowers surprised his colleagues and assorted officialdom by suggesting that the device that would do away with the need for the mechanical switching units should be constructed with thermionic valves. Many thought valves to be too fragile and too unreliable, but Flowers knew better. His work with the technology before the war had highlighted their shortcomings, but had also allowed him to minimise the effect of these.

In 1956, at the behest of the Post Office, the Joint Electronic Research Council was formed to work on a fully digital exchange. This consisted of the Post Office itself, plus five of Britain's six main telephone manufacturers – Siemens Brothers & Co, Automatic Telephone & Electric Company Ltd, Ericsson Telephones Ltd, the General Electric Company (GEC), and Standard Telephones and Cables. In 1962 a prototype 800-line exchange was brought into service at the Post Office's Highgate Woods facility in North London.

While the new technology was judged to have worked reasonably satisfactorily, the experimental exchange itself was deemed a failure – it was too large and complicated, and it soon became apparent that the newly invented transistors (that were used to perform certain functions alongside the valves) were both too expensive and not up to the task. (The system had a total of 500 valves, 5,000 transistors and approximately 400,000 electronic components.)[15]

The Post Office backed away from the all-digital development and chose to adopt older technologies, such as crossbar. Arguably, it should have revisited the design once appropriate quality components were available, but it didn't. By the time fully digital switches eventually came to Britain in the late 1970s, the technological lead the country had enjoyed due to the wartime work at Bletchley Park had evaporated. Moreover, the 1970s exchange, the so-called System X, was beset by problems: it was a four-way collaboration between the Post Office and its three remaining equipment suppliers (GEC Telecommunications, the Plessey Company and Standard Telephones and Cables), and, inevitably, it needed to interoperate with a network that by then contained crossbar and reed-relay (TXE) exchanges as well as the long-serving Strowger switches.

Having been the world's largest exporter of telecommunications equipment before the war and the technological leader after, Britain fell down the rankings year after year throughout the 1950s, '60s and '70s, to the point where such sales were so few and far between as to be almost meaningless.

In the post-war period great strides were made in several key technologies that are central to today's telecommunications industry. The first took place on 17 June 1946, in St Louis, Missouri, when the Southwestern Bell division of AT&T launched the first car-phone service.

While this represents a major milestone, this early manifestation of the mobile telephone had several disadvantages. First, because the equipment used valve technology, it weighed about 35 kilograms and took up most of the space in the car's boot; second, the service cost a lot, with the monthly

---

15 In December 1947 Walter Brattain and HR Moore of Bell Labs had demonstrated a working transistor, and in 1956 Brattain, together with co-workers William Shockley and John Bardeen, were awarded the Nobel Prize for Physics for this discovery. However, in 1925 Julius Edgar Lilienfeld, a physicist from Austria-Hungary working in Canada, had been awarded a patent for a transistor (although there is no evidence that he actually made one), and a German by the name of Oskar Heil patented a 'field-effect transistor' in 1934.

subscription set at $15 and call charges of 30–40c for a local call (about $205 and $4.15–$5.50, respectively, in present-day values); and third, capacity on the network was very limited, as the amount of bandwidth made available for the service was strictly rationed. America's Federal Communications Commission (FCC) wasn't inclined to give the new service access to much of the available radio spectrum; this was, after all, a scarce and potentially valuable resource – and might it not be better used to provide TV?

Someone had already thought of a way of improving capacity, however: in 1945 EK Jett, the then-head of the FCC, mentioned the idea of 'small zone' (or cellular) systems and noted that by such an arrangement 'in each zone the ... frequencies will provide 70 to 100 different channels, half of which may be used simultaneously in the same area without overlapping'.[16] Bell Labs put a lot of thought into the cellular concept, and in 1947 two of its engineers proposed a hexagonal structure for the cells.[17] Bell also considered whether such a system would be a commercial success – and concluded it would not. The technology just didn't exist, and so the cellular concept would remain nothing more than an impractical idea for the next 30 years.

Transistors would come to replace vacuum tubes, over which they had several key advantages, being far smaller, cheaper, more robust, generating less heat and having a longer working life. They would take a giant leap into the future in the 1960s, when the first commercial integrated circuits arrived. An integrated circuit is an electrical circuit with all of the components integrated on a single common surface. Various pioneers had suggested the possibility of such a device as far back as 1949, but the most significant developments had to wait until the best part of a decade later, when Jack Kilby at Texas Instruments produced a working example in September 1958. A few months later, Robert Noyce at Fairchild Semiconductor produce a more practical version of the same thing, using silicon as the substrate rather than Kilby's choice of germanium.

Gordon Moore, one of the early pioneers of integrated circuits, noted that the number of transistors on an integrated circuit of the same size

---

16 As reported in a 1945 edition of the *Saturday Evening Post*.
17 The hexagonal structure is more figurative than actual. The important aspect of the 'hexagon' is that adjacent cells all use different parts of the available radio spectrum, to avoid interference.

roughly doubles every two years – the famous 'Moore's Law'. This increase in the density of the components on a silicon chip had several benign consequences. Not only were the overall capacity and capability of the device increased but because the various elements in the circuit were physically closer together, the amount of time it took a signal to move from one to the next was reduced, thereby increasing the overall speed of the device. Power consumption went down, as did cost.

The development of ever-more powerful integrated circuits over the last 50 years is one of the key drivers of today's telecommunications industry. Smaller, cheaper chips have made it possible to produce devices of all kinds at prices that would previously have been unfeasible. A typical early integrated circuit might have featured a couple of dozen components that could perform a single, simple operation. Today's densest devices might have as many as 25 million transistors per square millimetre, separated from each other by a few billionths of a metre, capable of who knows what?

Fibre-optics technology also played a vital role in the development of telecommunications. The Romans were the first to learn that if heated, glass could be drawn out into fibres so thin that they became flexible and would conduct light from one end to the other. The idea that the light might be guided by refraction (a change of direction of a wave passing from one medium to another) was demonstrated in Paris in the 1840s, and the Irish physicist John Tyndall built on this.

Both John Logie Baird, the Scottish engineer who invented the television, and the American inventor Clarence Hansell demonstrated the transmission of images through glass tubes in the 1920s, while in the following decade the German scientist Heinrich Lamm demonstrated how bundles of fibres could be used for medical purposes – he invented the first endoscope (a tubular instrument used to look deep into the body) using optical fibres.

In the post-war years scientists such as Bram van Heel, Harold Hopkins and Narinder Singh Kapany worked on the transmission of images through bundles of fibres, but while this work expanded the science, it did little to further any significant commercial application for fibre. That situation changed in 1965, when Charles Kao and George Hockham at Standard Telephone and Cables' UK research laboratory realised that the high attenuation in existing fibres (which limited the range and the quality of the signal, rendering it essentially impractical) was not the result of some immutable physical law, but the consequence of impurities in the material.[18]

Kao and Hockham's work broke the back of the problem of attenuation and engineering did the rest. The American company Corning Glass made most of the early running, producing ever-higher-quality material with ever-lower attenuation, to the point where a fibre-optic cable wasn't only comparable to the best coaxial copper cables, but far superior. By 1977, when the first commercial fibre cable was laid (in Turin, Italy), fibre cables were capable of transmitting far greater bandwidths than copper, over longer distances, before (the inevitable) attenuation meant that reamplification was required. Fibres of that vintage could manage anything up to 150 kilometres, compared to copper's limit of 30 kilometres, and they were also cheaper to manufacture.

Over the 40 or more years since, the capability of fibre systems has improved almost beyond belief. Much of the best work has been undertaken by NTT Corporation, Japan's largest telecommunications company. In 2006 it set a new record when it managed to transmit a signal at a rate of 14 Terabits per second (Tbps) (14,000,000,000,000bps) over a 160-kilometre fibre. Bell Labs bettered this in 2009, with a 15.5Tbits transmission over 7,000 kilometres. A year later NTT was back in the record books with 69.1Tbits over 240 kilometres, and two years later it had raised the rate to 1 Peta bits per second (Pbps) over a 50-kilometre fibre. By 2017 the distance such a signal could travel unamplified had quadrupled to 200 kilometres. (These references to Tbits and Pbits are somewhat abstract for most people, but in simple terms a fibre with a 1Pbps capacity could transmit 10,000 hours of high-definition television video in a single second.)

On 25 September 1956 the first transatlantic submarine telephone cable opened for business, linking Gallanach Bay in Scotland with Clarenville in Newfoundland. The cable, named TAT-1 (Transatlantic No. 1), was a

---

18 Kao was awarded the Nobel prize for this work, somewhat belatedly, in 2009.

collaborative effort between AT&T (50%), the British Post Office (40%) and the Canadian Overseas Telecommunications Corporation (10%), and it cost about £120 million ($480 million) to build and lay. It began life with a capacity of 36 channels (enabling 36 simultaneous voice calls to be made). Together, on its first day alone, these carried nearly 600 calls between the UK and the US and more than 100 from the UK to Canada.

Building the cable had been far from simple. The proposed route had had to be surveyed to determine the degree to which the cable needed to be armoured to protect it from adverse conditions on the seabed, while the analogue signals that would be transmitted over the system's coaxial copper cables needed amplifying every few kilometres, so repeaters (signal boosters) had to be designed and prepared for installation in the cable. As the cable has a design life of at least 25 years, the components that went into making up these units had to be of the highest quality, as no one wanted to raise a 2,400-kilometre-long cable from the depths just to replace a faulty diode.

Once built, the cable and electronics had been loaded on to a special-purpose cable-laying ship and the process of wheeling out the cable began. The ship moved steadily, at about one knot per hour, all day, every day, for seven or eight weeks, unreeling the cable and its repeaters, and lowering them slowly on to the ocean floor.

However, as this was copper cable, it wasn't long before the errors inherent in such systems began to be of sufficient magnitude that they started to compromise the performance of the system as a whole. The on-board team of electricians would therefore be required to take preventative measures. First, the team had to measure the errors accurately. Next, it had to design special-purpose circuitry to eliminate those errors. This circuitry then had to be hand built and tested, and finally integrated into the next repeater in time for that device to be lowered over the rear of the ship some short while later. The process then began all over again the next day.

Copper cables such as this remained in production well into the 1980s, before eventually being replaced by fibre-based systems. Demand for international calls – especially across the Atlantic – continued to rise throughout the 1950s, '60s and '70s, and over the years TAT-1 was joined by

TAT-2, TAT-3 and so on, until TAT-7, the last copper cable, was laid in 1983. This had capacity for 4,000 simultaneous calls, or more than 100 times the capacity of the first in the series.

Laid in 1988, TAT-8 was the first transatlantic submarine cable to use fibre-optic technology. It was commissioned by a consortium of AT&T, British Telecom (BT) and France Telecom. Manufacture was undertaken by AT&T, Standard Telephones and Cables, and Alcatel, and it had ten times the capacity of its immediate predecessor (40,000 telephone channels or 280Mbps of data). Although fibre had been in use in terrestrial transmission systems for 11 years by then, the world's telephone companies had avoided deploying it under water because none of the manufacturers was prepared to guarantee that their optical repeaters would last for the duration of the cable's 25-year life.

TAT-8 wasn't as reliable as had been hoped, not through any fault of the components but thanks to the behaviour of sharks in its vicinity. The fibre system didn't use electrical shielding, as unlike copper, fibre is immune to electrical interference; the fish, however, were able to detect the high-voltage current running through the cable, which sent them into a feeding frenzy. Attacks on the cable invariably ended badly for all concerned: the sharks were electrocuted and the cable stopped working. Subsequent cables restored the interference shielding, after which shark and system coexisted happily.

Shortly after TAT-1 went into service, another far more dazzling event took place that was to have profound implications for communications and humanity. On 5 October 1957 the world awoke to the news that the Soviet Union had launched an artificial satellite into orbit. The spacecraft, Sputnik 1 or Prosteyshiy Sputnik-1 or Elementary Satellite 1, was giving out an intermittent radio signal. We take space-based communications for granted now but this is where it all began.

The world watched in wonder (literally – the satellite was clearly visible with the naked eye from many locations on the Earth's surface) until the spacecraft burnt up on re-entering the Earth's atmosphere on 4 January 1958.

Sputnik was certainly elementary in its design and ambition. Its onboard transmitter broadcast a series of bleeps back down to Earth, but it had no capability to transmit a received signal. The message that the Soviet Union wanted to get across to the rest of the world had little to do with intercontinental chat, and much more to do with its capabilities in the area of rocketry. This kickstarted what we now call 'the space race' and the American response was colossal and unambiguous: a new body was formed to oversee all work in the field – the National Aeronautics and Space Administration (NASA) – and by 1958 it had launched a lunar probe, Pioneer 1. Although this failed in its mission to reach the moon, it did demonstrate that communications could be transmitted through space, first from Cape Canaveral to the probe, which then retransmitted the signal to Jodrell Bank outside Manchester in the UK, then back to the probe, and on again in the same way to Hawaii and then Cape Canaveral, and finally, once again from Hawaii, halfway across the world to Manchester.

The end of the 1950s and early '60s saw further development efforts from the Americans and their allies, which culminated on 10 July 1962 with the launch of Telstar, the world's first communications satellite. On the day it was launched, it broadcast high-definition television signals across the Atlantic. Telstar was the result of international collaboration: the satellite itself was owned by AT&T, but the American company had called in NASA, its own Bell Labs facility, the British General Post Office and the French national post, telephone and telegraph (PTT) to help with the project.

The idea that a satellite could be positioned above the Earth in geo-stationary orbit (following the direction of the Earth's rotation so that it appears motionless at a fixed position in the sky) dates back to 1945, when the British science-fiction writer Arthur C Clarke proposed the idea in a magazine article. At a certain point (35,786 kilometres) above the Earth, a satellite will orbit the Earth at the same rate that the Earth spins on its axis, thus giving observers on the ground the impression that it is staying in the same place.[19]

---

19 In the past, communications satellites of this kind were routinely used to provide telephone services, but two-way voice poses a problem in that it introduces fractional delays into a conversation, as the signal bounces off the satellite and back down to Earth. This isn't an issue when transmitting most other types of signal, so today the majority of the traffic these satellites handle is either data or video.

There are several problems with the use of satellites for voice traffic. The first is cost. Building a geostationary satellite is an extremely complex task, requiring precision engineering of the highest order. Satellites need continuous power, which is provided by huge arrays of solar panels. To maximise the amount of sunlight these panels absorb, they have to turn through 360 degrees once a day, every day, throughout their 10+-year life. No lubricants are possible as these would just escape into space, so the machining of the mechanism has to be impeccable.

The launch vehicles are also expensive, as is the launch itself. This hazardous process occasionally ends in failure, so launch insurance is essential – but adds a considerable amount to the overall cost. Once in space, the position of the satellite has to be optimised to provide the best-quality signal over the designated coverage area, while at the same time minimising the need to expend fuel keeping the vehicle in its correct orbital location. There are only a limited number of these and they have all been taken, the vast majority being controlled by a handful of large international satellite companies, such as Intelsat, Inmarsat, SES Global and Eutelsat.

Finally, two-way voice poses a problem, in that it introduces fractional delays into a conversation, as the signal bounces off the satellite and back down to earth. This is not an issue when transmitting most other types of signal, so today the majority of the traffic they handle is either data or video. A new generation of geostationary satellites has been developed to carry very high throughputs (multi terabit) in the Ku and Ka bands, and these are now being supplemented by new medium and low earth orbit satellites in orbit at around 550 kilometres. (This is a topic we will touch upon in greater detail later.)

CHAPTER 3

Going mobile

The idea of a portable telephone is almost as old as the telephone itself. References to the concept date back to at least the early 1900s, while the German train service Deutsche Reichsbahn offered first-class travellers a telephone service on the route between Berlin and Hamburg. In 1946 comic-strip detective Dick Tracy used a two-way wrist radio – a fictional precursor to not only the mobile phone but the smartwatch too. And in 1996 the American TV series *Star Trek* took the concept a stage further.

But it's one thing to envisage something and quite another to make it a reality. This progress from the *idea* of a mobile phone to fact of its existence is, arguably, one of the more significant triumphs of 20th-century technology.

Portable radio sets were used by the military on both sides in the Second World War. Cumbersome, with limited range, short battery life and mediocre sound quality, these were a long way away from the sleek handsets of today, but improvements were continuous. In the post-war years, inventors in various countries miniaturised and produced hand-portable handsets. Limited to the range of a single transmitter, these 'cordless phones' lacked the capability for call-handoff (the process of transferring an ongoing call or data session from one channel connected to the core network to another channel) that is an essential feature of today's cellular phones. That key feature, the ability to move a call from one cell to the next seamlessly and without interruption, wasn't to materialise until the 1970s.

On 3 April 1973 Dr Martin Cooper of Motorola in the USA made a call to Dr Joel Engel, his rival at Bell Labs, using the world's first hand-portable cellular radio telephone. Cooper's phone weighed just over a kilogram, was 23 centimetres long, 13 centimetres deep and 5 centimetres across – but it worked. It didn't work for long, as battery life was limited to 30 minutes, but it was a start. Cooper later confessed that he'd been inspired by Dick Tracy's wristwatch phone.

By the mid-1970s, almost all the pieces were in place for the revolution in communications that was to begin in 1979. The handset had been made. The telephone networks that would carry the calls were moving swiftly from analogue to digital, while transmission systems had a new, exciting technology, which would massively boost capacity. Silicon had made more or less anything possible, and anything that was possible, affordable.

Before governments and administrations around the world were prepared to license a mobile-phone service, however, they had to be persuaded that such a system was needed. And if mobile-phone services were to be licensed, should it be on a monopoly basis or might it be better if the market were opened to competition? If the monopoly option were chosen, the question of which organisation ought to run that monopoly was easily answered – it was clearly something the PTT ought to address. But if the service were to be opened to competition with two or even three operators, how would those other operators be chosen? Would the PTT still be a natural choice for one of the licences? If it were, would it be able to bring all of the advantages of incumbency to bear upon the new market? Was this desirable? And was it appropriate that the state, through its ownership of the PTT, should be seen to be involved with one of the competitors, but not the others?

These were not easy questions to answer and it took some governments years to reach a decision.

For many, it seemed that the telephone industry was a 'natural monopoly' – that one entity could provide the service more efficiently and more cheaply than two or more suppliers. The idea has considerable merit when considering fixed networks with their huge embedded costs and assets but it begins to lose its currency if the network is not fixed but mobile. Yes, mobile networks also require significant initial investment, but any financial inefficiencies occasioned by overlapping coverage are likely to be more than offset by the benign effects of open competition.

In the United States, the ground had been prepared for competition in telecommunications by a series of decisions taken by the FCC in the late 1960s. Carter Electronics, a small electronics company, had developed a device that allowed someone using a two-way radio (such as those used by delivery companies and taxis) to connect to a third party through the public-telephone system. The Carterfone was an imaginative piece of engineering: the radio user would contact the controller and ask him to phone someone; the controller would make the call, then put the phone handset into a cradle that contained a two-way radio. A voice-activated switch toggled the Carterfone between 'receive' and 'transmit' mode, relaying the two-way radio's signal down the phone line.

The FCC liked the idea. As the connection to AT&T's network was acoustic, not electronic, there was no possibility that the device could damage the network, so in 1968 the FCC gave the Carterfone its blessing. This unilateral decision threatened AT&T's ability to determine what kind of equipment could be connected to its network, so it appealed against the FCC ruling in the US Court of Appeals. AT&T lost both the case and at the same time its monopoly on the supply of subscriber equipment.

This landmark ruling was noted far and wide, way beyond the confines of the US, and it provided encouragement to equipment suppliers the world over. While few of the leading PTTs were in both the manufacturing and operating game in the way that AT&T was, most nonetheless required all subscriber equipment to undergo a vetting process that many would-be vendors suggested was little more than a bureaucratic roadblock preventing change. Most markets, after the PTT lost its right to exercise this control, saw a sharp increase in the range and quality of available devices, accompanied by a reduction in their cost. Carterfone provided the breach in AT&T's armour that its competitors had long longed for.

It was soon to be followed by a far more significant breach.

In October 1963 a small company was set up, with not much by way of resources, but with plans to apply for licences to build a series of microwave relay stations between Chicago and St Louis. These would then provide long-distance connectivity to users of two-way radios travelling down America's famous Route 66. The name of this company was unambiguous – Microwave Communications Inc (MCI).

MCI's application for licences was granted in August 1969, allowing the company to provide a range of services over its growing microwave network, including data transmission and long-distance voice telephony. Around the same time, MCI gained a new investor and board member, William G McGowan, a man of considerable talent and great energy who immediately began applying for similar licences between other pairs of cities. Within 18 months he had created companies to operate a further 15 sets of links. After raising private equity in 1971, McGowan took the company public in June 1972, offering 3.3 million shares in the business at $10 per share, which valued the new business at an impressive $120 million.

It soon became clear to AT&T that MCI represented a small but significant threat to its monopoly long-distance business. Some of AT&T's local operating companies started making it difficult for MCI to achieve satisfactory interconnection terms between the local and microwave networks, and one, Illinois Bell, actually went as far as to disconnect MCI. If this move was meant to discourage McGowan, however, it had exactly the opposite effect: MCI filed an antitrust suit against AT&T in 1974 and moved the company's headquarters from New York State to Washington, to be closer to the seat of government.

The chances of the company winning its suit were hugely strengthened when later that year the Department of Justice joined in, filing another suit against AT&T that called for the corporation – the world's largest corporation by value at that time – to be dismembered, in the same way that Standard Oil had been, some 60 years earlier.

For the best part of a decade the lawsuits went on and on, but finally, in 1982, AT&T announced that the company had voluntarily agreed to break up, separating manufacturing and long-distance operations from those of the local exchange companies. AT&T had two years to finalise the details, with the proposed restructuring set to come into effect on 1 January 1984.

There's probably never a good time to pull yourself to pieces, but that period, from 1982 to 1984, was among the worst possible for those within AT&T who had an interest in the country's new mobile-phone licences. The FCC had been presented with a proposal as to how best to license the new technology in 1971, and eight years later it had reached a decision. Two years after that, in 1981, the process was, at last, under way. However, AT&T was distracted by all the legal stuff and probably didn't think hard enough about the commercial potential of the technology it had helped created.

On 1 January 1984, the old 'Bell System' re-emerged as the new Bell System, no longer one single entity but eight separate companies, all with a common heritage and similar outlook.

The US Department of Justice had determined that none of the new businesses should be larger than GTE, at that time the largest independent. To avoid that possibility, the Bell System had had to be split into at least seven separate companies. These so-called 'Baby Bells', all about the same size, had very distinct geographic and economic characteristics: NYNEX (the New York and New England Exchange Company), Bell Atlantic (Eastern Seaboard), BellSouth (most of the southern states), American Information Technologies Corp (Ameritech, the Midwest), Southwestern Bell (Texas and the rest of the south), US West (14 states from the Dakotas across to the west coast) and Pacific Telesis (California and Nevada). The long-distance business retained the AT&T name and ownership of Western Electric, the manufacturing arm, which had renamed itself AT&T Technologies.

The dividing lines between the new businesses were set: AT&T couldn't enter the local exchange market and the regionals couldn't enter the long-distance market. Although the long-distance company retained the iconic AT&T brand, it wasn't allowed to use the name 'Bell'. It also lost ownership of the lucrative *Yellow Pages* commercial directory business, which passed to the regionals. And the restriction on AT&T's entry into the computer business, which had been in place since 1956 (when the spectre of antitrust action had made it agree not to move into this field), was lifted.

The regional companies, rather than their one-time parent, were given the right to apply for the new cellular licences. According to some sources, AT&T had been offered the mobile businesses, but turned them down on the grounds that 'it'll never be more than a marginal thing'. This was something of a misjudgement, but AT&T wasn't alone in drawing this conclusion.

The award of the first US mobile licences wasn't a straightforward process and several mistakes were made. However, the FCC got one thing absolutely right – it set a common technology standard that all operators were required to use. In this case it was AT&T's Advanced Mobile Phone Service (AMPS), an analogue or 'first-generation' cellular design. Next, it made the market competitive by dividing the available spectrum into two equal blocks, designated 'A' and 'B'. Competition was good, though in this case it was on a somewhat unequal basis. The B block was to be given to the local phone company, the 'wireline operator', while the A block would be awarded after an auction to a so-called 'non-wireline operator' – a new entrant into the telephone market.

The FCC decided that cellular phone numbers should use the same area-code number format used by landline operators. There was probably a good reason for this but it gave rise to an immediate problem. Cellular phone charges were likely to be substantially higher than ordinary landline rates, which in many cases were free; a caller could unwittingly incur a heavy charge on a call they'd thought would cost them nothing. It followed from this that the only equitable approach was for the mobile-phone owner to pay to receive a call, as well as to make one. For years afterwards, until charges dropped to a more reasonable level, many US phone owners only turned on their mobile phone when they specifically wanted to make a call. This lessened the possibility of having to pay for incoming junk calls, but at the same time it severely diminished the utility of the service.

Finally, the FCC chose to split the country into a large number of separate areas, each of which would have its own licences. The country was cut up into no fewer than 734 cellular market area or CMAs, which were in turn separated into 306 urban regions known as metropolitan statistical areas (MSAs) and a further 428 rural statistical areas (RSAs).

There were so many things wrong with this arrangement. First, it was an inefficient and wasteful use of radio spectrum. Then, unless you were also the operator of the adjacent CMA or CMAs, you couldn't guarantee your customers the same standard of service once they left your coverage area and began 'roaming' on a neighbouring network – coverage may be poor or there may be no service at all. Third, fragmenting the market in this way didn't help operators achieve economies of scale – especially those operators that were new entrants. Fourth, there were issues arising from the use of different vendors – while AMPS was a standard, Ericsson's version of the standard might not be quite the same as Motorola's or AT&T's. And, finally, the need to allocate 1,456 separate licences created an administrative nightmare, as the FCC received tens, if not hundreds, of thousands of applications, each of which had to be separately assessed.

The B-block licences weren't really the problem. Awarding these was a comparatively straightforward process, as most of the time this meant giving the licence to the local Bell Company. In those instances where there was more than one operator in the franchise area, the wireline companies generally came to some kind of joint-venture arrangement, with proportionate equity stakes – not a matter to concern the FCC.

The A-block licences were quite another matter. Even though the FCC had decided to license the MSAs in tranches (starting with the 30 largest, in descending order of population, from New York City down to Hartford, Connecticut), the task was immense and the process soon got out of control. With each successive tranche of licences, 50 or even 100 bids for the same franchise weren't uncommon.

In the end, the FCC gave up, and in 1984 it called on an obscure piece of legislation that allowed it to offer the licences by way of a lottery.

It took the best part of a decade for some of the smaller licences to be awarded, by which time the B-block operators had been up and running for quite a while – and had secured all the most attractive customers. For example, in New York, NYNEX launched its mobile service in June 1984 and enjoyed an effective monopoly for about two years until Metro One started up in April 1986; and in Los Angeles, PacTel introduced a cellular service in June 1984 (just in time for the Olympic Games) but the Los Angeles Cellular Telephone Company didn't get started until March 1987, nearly three years later.

Over the coming years, most of the rest of the world looked at the licensing fiasco in the US and resolved not to go the same route. Nowadays, regional licences are a feature in only a handful of countries – the USA, Argentina, Brazil, Canada, Colombia, India, Indonesia, Japan, Mexico, Peru, Russia and Iraq.

As the dust began to settle after the initial US licence awards, eight large mobile-telephone operators emerged – one was owned by GTE, the largest independent operator, while the seven others were the mobile subsidiaries of NYNEX, Bell Atlantic, BellSouth, Ameritech, Southwestern Bell, US West and PacTel. The footprint of each was defined by its parent's wireline presence, as one might imagine – after all the exertions of the licensing process, the FCC had managed to create a mobile landscape that looked remarkably like the prevailing local landline landscape.

This was to change quite soon, though, as some of the new, non-wireline operators began to expand, which they did mainly through the acquisition of their counterparts, the other A-block businesses.

Ameritech was the first of the seven Bell regionals to launch its mobile service, in October 1983 in the city where it had its headquarters, Chicago, Illinois.

This made it the first in the USA but not in the world: that honour went to the Japanese company NTT, which had launched a mobile service in central Tokyo at the end of 1979. For over a year, this had the distinction of being the only cellular service in existence, but then, in quick succession, it was joined by networks in Sweden (October 1981), Norway (November 1981), Finland (March 1992) and Denmark (September 1982). As with conventional telephony, Sweden stood apart from other markets and licensed two separate operators, one controlled by Televerket (today's Telia), the other by an independent entity called Comvik. The other three countries awarded their licences to their state-owned PTT.

All these networks were based on analogue technologies. Japan used a system called Nippon Automatic Mobile Telephone System (NAMTS) developed by NTT. The Comvik network in Sweden used a proprietary technology also called Comvik, while the Swedish PTT opted for the Nordic Mobile Telephone (NMT) standard developed by LM Ericsson. The three other Scandinavian countries also selected NMT, in a rare early example of regional solidarity.[20]

With the launch of the US cellular network in 1983, the total number of markets with functioning networks worldwide amounted to six (or seven, if you count the private NMT network launched in Saudi Arabia to keep the royal family in touch with each other). More followed in 1984, as the technology spread across the world. AMPS was preferred in most markets, and during 1983 four networks based on the US standard were brought into service in Hong Kong, Indonesia, South Korea and the United Arab Emirates (UAE). Just before the end of the year Austria joined the party, when the PTT's Mobilkom subsidiary launched a mobile service using a standard called C-450.[21]

In 1985 the pace picked up, with ten new markets opening: the UK and Ireland based their systems on TACS; France launched a system called RC-2000; Italy chose a design that had been developed in-house called Radio Telefono Mobile di Seconda generazione (RTMS); the Netherlands, Malaysia, Oman and Tunisia opted for NMT; West Germany adopted the same C-450 design that the Austrians had just launched; and Canada fell in line with the US. All bar the UK and Canada were monopolies, with the licence given to the local phone company.

In all, cellular-based mobile-phone systems were now in place in 23 markets around the world, based on seven separate and incompatible designs.

The year 1985 was notable for another first: two further licences were granted in Hong Kong, to privately owned companies controlled by Hutchison Whampoa (now CK Hutchison) and Millicom, a small but

---

20 Actually, the choice was easy: NMT was initially designed to operate at around 450MHz, compared to 800MHz for AMPS and 900MHz for its subsequent UK variant, Total Access Communication System (TACS). Physics dictates that the lower the frequency, the further a signal of any given strength will travel. The trade-off is that the network capacity is lower for any given bandwidth – but that, at such an early stage of the market's development in countries with large landmasses and low population densities, was essentially irrelevant.

21 A variant of the Scandinavian NMT-450, German engineers had modified it to improve certain aspects of performance, at the cost of incompatibility with other NMT systems.

aggressively expansionist American company. This interesting move made Hong Kong the first place in the world to have three-way competition in mobile communications.

The launch of mobile services in the UK represented Vodafone's first steps on to the international stage, but it was only one part of a much broader plan to liberalise the entire telecommunications market in the country by introducing competition and new technologies and services, including cable TV and mobile telephony. Although greeted with scepticism by politicians and operators in almost every market where the telephone business was controlled by a state-owned monopoly, the move was to prove very successful and in the following years was emulated worldwide.

The general election in the UK in May 1979 saw the Conservative Party ascend to power, with a mandate for change. Income taxes were slashed, foreign-exchange controls were abolished, and the power of the trades union was constrained. Margaret Thatcher's party also had radical plans for revitalising industry, the two most important for telecommunications being the widespread adoption of free-market policies and the privatisation of state-owned industries.

The government knew that the telecommunications market badly needed reforming, and over the course of the next 18 months several important strides were taken towards that goal. First, on 1 October 1981, British Telecom (BT) became a public corporation in its own right, separate from the Post Office. The newly independent BT enjoyed a monopoly on everything to do with telecommunications in the UK.[22] While this was a necessary first move, it wasn't any kind of solution to the problems facing the company. Its equipment was out of date and unreliable. The long-awaited transformation of the switching system from analogue to digital had begun but very few lines were served by the new technology. Trunk (long-distance) transmission systems were mostly copper-based, though microwave links were also used, with a few of the more remote

---

22 Except in the Borough of Kingston-upon-Hull, an anomaly that was a result of an obscure loophole in the 1911 nationalisation process.

long-distance connections (such as those to the Scottish islands and North Sea oil rigs) being provided by such exotic technologies as tropospheric scatter. Almost all the telephone handsets in use were of the old-fashioned rotary-dial design and these were hard-wired into the home and immovable once they'd been installed, as it was illegal to alter the wiring of any item of equipment BT had supplied – and of course, they had supplied it all.

But the state of the network wasn't the biggest problem facing BT in the early 1980s. Its bloated workforce of nearly a quarter of a million employees drained profit out of the business without providing much in the way of public service. Restrictive practices were rife, as was inefficiency and incompetence. Applicants for new phones might have to wait six months or more, during which time they would earnestly hope that both the indoor installation engineer and the outdoor specialist turned up on the same day, as one wouldn't work without the other. Many potential customers were left frustrated and unserved, and it wasn't uncommon for such people to resort to bribery. It was no wonder that fewer than half the homes in the country had a telephone. In the years to follow, BT's management addressed the issue of the workforce but it was neither a quick nor an easy process.

The next important government initiative was to create a regulatory body to oversee the newly commercialised entity. The Office of Telecommunications (OFTEL, now OFCOM) was created in 1981, headed by eminent academic Professor Bryan Carsberg.[23] To ensure that the new body was able to perform effectively, the government granted it considerable powers, including the ability to limit the prices BT charged its customers.

The third important initiative was to introduce a competitor, so an invitation to apply for a second UK licence was published. Extraordinarily, only one applicant came forward: Mercury Communications, a consortium

---

23 OFTEL has had several heads, but none has matched the professor. He had a very clear understanding of his role, which he said was to create a genuinely competitive market, at which point his services would no longer be needed. Later regulators in the UK and elsewhere seem to have forgotten this simple imperative.

owned by the newly privatised Cable & Wireless,[24] the oil company BP and Barclays Bank, which duly received this second licence in February 1982 and immediately began constructing a long-distance figure-of-eight fibre network connecting Mercury's headquarters in Birmingham with London, Bristol, Nottingham, Manchester and Leeds.[25]

Impatient to get competition into the market as quickly as possible, the government decided to remove BT's monopoly on equipment located in customers' premises – the same thing that had happened to AT&T in the USA some 14 years earlier. At the same time, the market for value-added services (which, according to OFTEL, was anything that wasn't basic telephony) was opened up.

Finally, on 19 July 1982, the government announced that it would sell up to 50.2% of its interest in BT to the public. The first great privatisation of modern times began in November 1984, when the government offered 3,012 million ordinary shares in BT to private investors (out of a total of 6,000 million). The offer was 3.2 times oversubscribed, meaning that all applications were scaled back.

The share offer contained a clever element to attract private investors: although the issue price was 130p per share, only 50p was due on application, with the balance spread across two further payments some months later. So when shares began trading at 80p on the morning of 3 December, smaller investors had an immediate 60% gain, which many cashed in immediately. Did this help fund Christmas celebrations up and down the country? It wasn't much remarked upon at the time, but someone noticed;

---

24 Cable & Wireless, a direct descendant of the 19th-century British cable companies, was actually the first company to be privatised in the UK: in November 1981, an initial tranche of 50% of the equity was made available to private investors at a price of 168p per share. The share sale was successful, despite widespread investor confusion about what the company actually did.

25 Thatcher and her principal industrial strategist, Sir Keith Joseph, had another motive for wanting additional competition, but this was never greatly aired in public. Both had been appalled by the 1973 miners' strike and the way it had brought an end to the last Conservative government. They blamed the unions for that, and also for the chaos that was the 1978 'winter of discontent' during which there were widespread strikes by public-sector trades union demanding ever larger pay rises … plus, of course, for almost everything else that was wrong with the UK's industry. Joseph noted that the vast majority of BT's employees belonged to either the National Communications Union or the Post Office Workers' Union and was concerned that the phone system was as vulnerable to strike action as the mining industry had been. The un-unionised Mercury was, in part, designed to be a safety net in the event of industrial action at BT.

trading in almost all of the main, subsequent privatisations was timed to begin in early December.

While all this was happening, other moves were afoot to create additional competition. Much had been heard about the Americans' decision to license cellular-radio (mobile-phone) networks, so it required no great leap of the imagination for the British to do the same thing. Thatcher liked the idea of competing networks, so the decision was taken to emulate the Americans and award two licences. Obviously, BT would have to be involved with one of them, but to whom should the second be given? Requests for applications went out, three teams put themselves forward, and the licence was awarded to a new company, Racal Millicom Ltd.

Two other aspects of the UK process are worth noting. First, although BT had effective control of the company that ran the first network, it wasn't initially allowed to own more than 60% of the equity. Nor was it allowed to use the BT name – though at this stage, that was probably no bad thing, as BT didn't have a great reputation. It chose to call the new company Cellnet. A partner also had to be found to take the 40% not allowed to BT; eventually one was found in the shape of the Securicor Group, a company with no credentials in the mobile business other than that it used traditional car-phones in its day-to-day security business. This wasn't particularly helpful to BT, but it wasn't meant to be.

Second, although the idea of two competing mobile networks was a step in the right direction, it wasn't quite enough for Mrs Thatcher. She wanted more competition in the new market. A third licence probably wasn't a good idea; on the basis of 'divide and rule', BT could more easily dominate a three-player market than it could a duopoly. There was a real fear of this, fuelled in part by BT itself. A spokesman for the company predicted a less-than-stellar future for the new business when he declared, 'There'll only be 20,000 customers by 1990 and we'll have 80% of them.'

There wasn't enough spectrum for a third operator anyway, so it was decided to separate the running of the network and the provision of services to the end customer. With this in mind, the UK licences prohibited the network operators from dealing directly with the end customer, this being the preserve of a new type of business, the service provider. Each operator was allowed to own just one service provider, which could not be in any way cross-subsidised. The thought was that the business of service provision would attract all sorts of entrepreneurially minded types and would help kickstart the new industry – an accurate prediction.

The first UK mobile-phone network, Racal's Vodafone, went live on 1 January 1985. Eight days later, its competitor, BT's Cellnet (now O2), was launched.

CHAPTER 4

Vodafone – the mobile industry's dynamo

A
T&T must take a large amount of the credit for the creation of the telephone industry. It's rightly recognised as the inventor of the technology, and over the years that followed it continued to develop and innovate until it eventually gave us the technology to make possible universal mobile communications.

However, the credit for realising that potential and transforming it into a global phenomenon lies mostly with other organisations and especially with Vodafone.

Vodafone's roots aren't those of a traditional telephone company. It emerged from an obscure British company, Racal Electronics,[26] which specialised in defence equipment and in particular military radio communications. But Racal wasn't really a technology business in the way that some of its UK contemporaries were; it didn't exist to innovate in the way that Plessey or Ferranti did. It existed to sell, and sell it did, all over the world: if it looked like there was trouble ahead in some distant land, Racal would have a salesman on the next plane. In the mid- to late 1970s its profits grew dramatically, from £6 million in 1974 to more than £100 million in 1981. It looked unstoppable.

But the rate at which new orders were coming through began to slow towards the end of the decade, and by 1982 the profit progression was

---

26 Racal's name derives from that of its two founders, Ray Brown and Calder Cunningham.

beginning to go into reverse. Attempts to diversify and develop new businesses – modems, computer-aided design and the like – enjoyed some success but by no means enough to offset the downturn in military radios.

During 1982, when the first Thatcher government in the UK announced the two mobile-phone licences, Racal bid for the second licence because its dynamic chief executive Ernie Harrison thought the new service would provide the company with an opportunity to sell more radios – not military radios, but commercial cellular radios, or mobile phones. Racal knew nothing about the telephone business and almost nothing of consumer markets, so it set up a joint venture with the US company Millicom, and Hambros Advanced Technology Trust, a fund that had been established by the venerable Hambros Bank.

When it came to bidding, the consortium led by Cable & Wireless – until the previous year owned by the British state – was the hot favourite. Cable & Wireless was also one of the three partners in Mercury Communications, the company that had been awarded the UK's second telephone licence and it had recruited two of the only British companies other than BT with any knowledge of telecommunications, Air Call (a paging company)[27] and Telephone Rentals (which specialised in renting large Private Automatic Branch Exchange (PABX) systems,[28] the one bit of the CPE market that had not been a BT monopoly).

The second team consisted of Ferranti (the company responsible for the world's first commercial computer), London Car Telephones (which had been an agent for BT's pre-cellular car-telephone service), and Graphic Scanning (which owned the largest pager company in the US and had recently won some of the first A-block cellular licences in that country).

Even with Millicom and Hambros on board, Racal looked to be outgunned. However, on the day of the pitch, it managed to strike all the right notes – after all, marketing was its forte. The enthusiastic Racal team stressed the growing importance of communications, the need for real

---

27 A pager (or beeper) is a device that receives and displays alphanumeric or voice messages. Pagers are still used by some emergency-service personnel such as doctors.

28 A PABX is a telephone system owned by a business that connects the business's instruments separately to the public telephone network; every internal call made is routed through it.

competition within the market, and the potential for exports of both equipment and service expertise. Astutely, it suggested that this new technology could help revolutionise business practices, a view that Thatcher shared. And it tore into BT's timid prediction that there would only be 20,000 subscribers by 1990; Racal said that the total market would be closer to 500,000 users and that it would have at least 50% of those on its network. (For what can't have been much more than a wild guess, Racal's forecast of the market wasn't bad – the actual figure at the end of 1990 was 1.15 million, with Racal controlling 57% of those.)

The three-man committee that had been brought together to assess the alternatives was headed by Professor Carsberg. His view – and that of his colleagues too – was that the Racal team seemed the most committed to and excited by the opportunity. Racal Millicom was duly awarded the country's second licence in May 1993.

Racal Millcom now had just under two years in which to design, build and launch its network if it were to meet the March 1995 deadline in its licence. Its partner could provide some technical help, as could its network supplier (assuming it chose wisely), but finding the money to build the network would be difficult and there was a huge amount to learn.

A comparison of BT and Racal gives some idea what the smaller company was up against. In its most recent financial year Racal had achieved sales of just over £800 million and made a profit of around £120 million before tax. BT, by contrast, had reported total revenues of over £7 billion and profits of over £1 billion – more than Racal's total turnover. Moreover, BT had a vast team of engineers and a research capability that rivalled that of Bell Labs. Cellnet had access to the entire BT landline network and could locate its masts in any or all of BT's thousands of exchange buildings up and down the country.

Looking back, it's difficult for us to imagine the difficulties Racal faced. These days, the process of building and launching a mobile network is well understood, as it's been done more than a thousand times by now. Back in 1985, however, everything was new and had to be worked out from first principles: Racal was staring at a blank canvas. It had to learn the rules of the new game – and make up a few of its own – as it went along.

Racal may have had no real idea about how to build a telephone network, but it realised that the choice of switching equipment was key. Examining the options, it selected Ericsson's AXE switch, on the grounds that it was the biggest and most powerful system available at the time.

Unlike BT, Racal didn't have access to large tracts of land on which to locate the system's base stations, and it had little or no presence in London, where many of its customers were likely to be. It had to adopt a highly pragmatic approach to site locations, making use of anywhere that came to hand.

And Racal needed to find people to run the new business. There was a strong pool of talent within Racal to draw on, but for Harrison the question of who to put in charge wasn't difficult to answer: his old friend and colleague Gerry Whent was the obvious choice. The 57-year-old knew his way around Racal; he'd held several different posts within the group over the previous 15 years, the most recent as managing director of Racal's Radio Communications Group. Harrison trusted Whent's judgement and knew he could rely on him.

Harrison and Whent agreed that some positions were probably better given to outsiders, and 36-year-old Chris Gent, who at the time was in a managerial position at ICL, the British computer company, was head-hunted as the managing director of the network business. Gent's natural charm made him easy to like, and his ability to get on with people of all shapes and sizes was to prove crucial on more than one occasion in the future.

Someone also had to be found to head the marketing team, and the highly motivated Julian Horn-Smith, a young executive at the Mars Company, was enlisted. A good judge of character and very shrewd, Horn-Smith was an excellent foil for Gent.

The new business's first finance director was Ken Hydon, who'd joined Racal in 1977 and had held various jobs within the group. Enthusiastic and optimistic by nature, he was to establish a strong rapport with Gent and Horn-Smith over the years they worked together. This triumvirate deserves a lot of the credit for making Vodafone the success it became.

The key roles having been filled, all that remained were to name the new business and to establish its headquarters. After the failure of the pre-launch prime-time television advertising campaign, 'Racal who? The biggest company you've never heard of ...', Saatchi and Saatchi advertising agency was called in and asked to come up with a name. This resulted in 'Vodafone', a portmanteau of 'voice' and 'data(f)one'. The new company located itself in the small country town of Newbury in Berkshire some 50 miles (80 kilometres) west of London.

One of the tasks needed to ensure the success of the venture, indeed, of the whole new industry, was to encourage independent dealers to market the service. The idea was that service providers would deal directly with the end-customer and be responsible for distribution. The proposed mechanism allowed them to acquire airtime on a wholesale basis, which they would then sell on to the subscriber at a retail rate. An additional source of revenue came from bonuses paid by the networks for each new connection they made.

It was a simple business model and it appealed to a lot of entrepreneurial types. Within weeks, numerous small companies had applied for licences to wholesale the new service, companies that might up until then have specialised in selling used cars or repairing televisions.

The degree to which the two British networks, Cellnet and Vodafone, used service providers varied significantly and was one of a number of key factors that helped influence the way the market was to develop in those crucial early months. Under the terms of their licences, both networks were allowed to own one in-house service provider. Cellnet wasn't attracted to these intermediaries; it couldn't see that they brought any value to the business.

Whent too had reservations at first, but suddenly, the penny dropped, and he realised that these organisations could extend Vodafone's reach – without the need for additional capital. He also recognised that in many ways the new service providers were analogous to the agents he'd used at the Radio Group to facilitate deals in the Middle East and elsewhere, so he decided they should be encouraged.

One of the key decisions facing the two new companies was how to price the new service. BT's existing phone tariffs were complex, with call charges varying according to distance and time of day, depending on whether the call was made during peak, standard or off-peak hours. Cellnet, it seemed, planned to replicate this approach within the new industry.

Whent, however, was determined to keep the business simple and prevent the introduction of this unnecessary complexity, so he pre-empted the issue by suddenly announcing Vodafone's proposed charges: there would be a £50 fee to join the network, a monthly subscription charge of £25 and a per-minute call charge of 25p. The simplicity of the tariff appealed to dealers and customers alike, and a few days later Cellnet announced an almost identical package.

Vodafone's 1985 prices may look expensive now but should be seen in context. Back in 1985, BT charged its residential customers £15.15 per month just to rent a telephone line, while the fee for business connections was £23.50. Call charges varied hugely, but were generally far higher than today's, with some international rates being almost unaffordable. And because Vodafone had settled on a fee structure that was well below what Cellnet had been thinking of charging, the market grew rapidly.

However, success was still some way away. While insiders could see the progress that was being made and the quality of the business that was being created, it was hard if not impossible for Racal to communicate this to the outside world. Throughout 1983 and '84, the company continued to come under pressure from its shareholders. The military-radio business was still not delivering, while the group's other divisions were finding the going increasingly difficult too; and on top of this, the new Vodafone business was incurring start-up losses. Several analysts predicted that cellular radio would be a disaster, and it took great courage on the part of Harrison and Whent to ignore these dissenting voices and continue pouring funds into the business.

There were other issues that needed to be addressed too. Vodafone had taken the lead on pricing and had signed up a number of independent service providers (ISPs), but as 1984 drew to a close, there was a growing awareness within the company that it was lagging behind in one crucial respect – it didn't have much of a retail-distribution network. BT had used its scale and vast national presence to help Cellnet sign up many more agents and retailers. Then suddenly, out of the blue, Horn-Smith signed up more than a dozen agents in just a few days, including the Automobile Association, one of the largest motoring organisations in the world.

'Hi, Dad. It's Mike. This is the first-ever call made on a UK commercial mobile network.'

This is how the first call made over the Vodafone network began, just after midnight on 1 January 1985. Michael Harrison, one of the CEO's sons, had gone to Trafalgar Square to join in the New Year's celebrations armed with a Vodafone VT1 mobile handset, and he'd used it to call his father.

A slightly more formal ceremony would take place a few days later to mark the launch. Vodafone was up and running a full three months before the date stipulated in its licence – and, just as important, eight days before Cellnet.

The handset Michael Harrison had used weighed something in the region of five kilograms and it wasn't cheap – but then, nor were any mobile phones back then. Vodafone's basic car telephone sold for £1,415 and top-end models could cost anything up to £3,000 – about £8,000 in today's money. This was an obstacle for most people. Nonetheless, by the end of the 1984/85 financial year three months later, Racal announced that it had connected 2,711 subscribers. It didn't quite match its competitor's figure of 4,500, but then the Cellnet numbers were augmented by a steady flow of customers from BT's pre-cellular System 3 and System 4 car-phone services.

As the weeks passed, the numbers on the new networks continued to increase. By June, Vodafone had added a further 4,000, while Cellnet had reached a total of 10,000. By the end of the two companies' 1985/86 financial year the following March, Vodafone had narrowed the gap between the two networks to just 3,000. It drew level with Cellnet three months later, and in July 1986 it took the lead.[29]

There were several factors behind Vodafone's victory in this all-important

---

29 It would be nearly 20 years before Racal would drop behind Cellnet again, in December 2005, by which time the number of subscribers on the network had become rather less important than the profits they generated. The two companies' UK operating profits for that year were £975 million for Vodafone, and £623 million for Cellnet.

early phase. First, there's no doubt that it benefited from BT's poor image. Although it had been part-privatised, BT was still at heart an old-school monopoly, with all that that implies. Cellnet itself was different, and better than that, but the public weren't dealing with Cellnet, they were dealing with its service provider, British Telecom Mobile Communications – so Cellnet became tarred with the BT brush.

A second factor was the two business's use of service providers. Vodafone understood that they were there to make money, so the best way to treat them was to ensure that they understood the intricacies of service and provided with all the necessary support. Cellnet preferred to hand out expensive freebies instead.

Third, there was the issue of network quality. Cellnet had opted for a highly scalable, distributed architecture based around Motorola switches, with radio links engineered in-house by BT, but in practice the system didn't live up to expectations. The switches spent most of the time chatting to each other, passing interesting data back and forth, but leaving little room on the system for actual revenue-generating traffic. And despite access to BT's vast property portfolio, Cellnet's coverage wasn't as extensive as Vodafone's – Cellnet had located its base stations inside or on top of local telephone exchanges even if those weren't the best places to put them.

All of these factors meant that customers on the Cellnet network made fewer calls than those on Vodafone's – and therefore generated less revenue. This meant a lesser share of the revenue for the service provider as well as the network operator, as, it should be remembered, the service providers were remunerated by reference to each subscriber's overall expenditure. Noticing this, the middle-men began steering prospective customers towards Vodafone. And as the percentage share of revenues that the service providers received increased in line with the number of customers they connected, it didn't take long for this process to become self-perpetuating.

Cellnet's network issue wasn't easily resolved and capacity continued to be a problem until eventually the company decided that it had to rethink its approach to the network, rebuilding it from the ground up. Stumbling like this at such a crucial early stage held back Cellnet's progress throughout most of 1986 and provided further impetus to Vodafone.

Vodafone deserved its lead – it had been more flexible, more responsive and more imaginative. By the end of 1986 the company was connecting over 1,000 customers a week, and by the end of December it had more than 63,000 subscribers – a lead of 9,000 over Cellnet. It had moved into profit on a day-to-day basis some months ahead of expectations. Racal felt able to defy critics and place another heavy bet on the new business.

Back in 1983, when the original consortium had been assembled, Millicom had taken a 15% share of the equity, and was also gifted a 10% share of all subsequent Vodafone profits in perpetuity. That deal was beginning to look quite absurdly generous but how could Racal do anything about it? It needed a bit of good luck – and it got it. Millicom, which was itself expanding into a number of new mobile markets, suddenly found itself presented with a large tax bill that needed to be paid by the end of the year. It asked Racal what it would offer for the 15% stake and, after some negotiating, a price of £69 million was agreed, together with a further £40 million in return for the cancellation of the royalty arrangement. A few days later, Racal also acquired the 5% stake owned by Hambros.

With the Americans' departure, Racal Millicom became Racal Telecom.

Some of Racal's institutional shareholders didn't think much of these arrangements: Racal had forked out just over £130 million for 20% of a business that hadn't yet reported a profit. But of course it was a brilliant deal – by the early years of this century, the saving on the royalty alone amounted to well over £2 billion.

The Vodafone licence allowed the company to offer other services in addition to operating a mobile-telephone network. There was the business of value-added services (basically, anything that wasn't straightforward telephony), paging and, of course, service provision. Vodafone adopted the same approach Racal used, creating separate companies for each of these activities. It fell to Chris Gent to manage the first and last of these, Vodata and Vodac, while Julian Horn-Smith was given responsibility for Vodapage and international development.

At the time Racal had pitched for the licence, it had made much of the opportunities it foresaw for manufacturing and selling cellular-related equipment – not just handsets but network equipment as well. Vodafone took over some of Racal's manufacturing facilities and began to develop its first hand-portable telephone. By the end of the year the first of these were coming off the production line. The unit was soon dubbed 'the brick', due to its size, shape and weight, perhaps revealing its military DNA.

The flirtation with manufacturing didn't last long. Racal realised it probably couldn't be both an equipment supplier and a network operator, because this might result in it competing for its own customers – it would have to choose between the two. It pondered for a while which path to take and chose wisely: it rolled the manufacturing activities into a joint venture with Plessey, named the business Orbitel, and pretty much left it to its own devices.[30]

This move allowed Racal to focus its attention on mobile services and in particular the Vodafone business. An unconventional analyst at Kleinwort Benson, a leading London merchant bank, noticed that US cellular companies were trading at prices that valued potential customers at anything up to $55 each – '$55 a pop' in the industry's vernacular. (The 'price per pop' was a key metric in the early years of the industry – businesses couldn't be valued by reference to their earnings, as they hadn't any yet, so some other yardstick had to be invented.) Applying that number to Vodafone gave a figure of slightly more than $3 billion (£1.6 billion). That was more than Racal's total capitalisation, for Vodafone alone! Something was out of kilter – and that was the Racal share price. Other City analysts began changing their minds and the company's valuation began moving upwards again, eventually reaching £3 per share – its old high-water mark not seen since 1982.

Then, all of a sudden, in October 1987, stock markets around the world began to plummet. In just a few weeks, Racal's shares fell from an early October high of 350p to about 200p. At that level the company was valued at just £1.3 billion. Share prices can go down as well as up, and that had

---

30 Shortly afterwards, Plessey was bought by its old rival GEC, after which the business was sold off to Ericsson.

happened here, with a vengeance.

In the months that followed the October 1987 crash, Racal's management watched the company's share price slowly creep up again to around 240p. At that level, the Racal Group was valued at just £1.5 billion.

They believed Vodafone alone was worth at least that much. This wasn't just corporate bluster – Harrison and his colleagues knew that the 1987/88 numbers were going to be way above even the most optimistic forecasts: the subscriber base had nearly doubled to 160,000, market share was up again, to 55%, and crucially, operating profits were up by some 400% year on year, to over £50 million. Management's earlier promise that Vodafone would achieve annual profits of £70 million before the end of the decade no longer looked outlandish but a certainty.

The share price was important because a higher share price made it easier to raise additional funds and less likely that any fund-raising would dilute earnings. The other side of this coin was that a low share price, which undervalued a business, left the company vulnerable to a possible hostile takeover, as Racal was soon to discover.

The rumours that Eric Sharp of Cable & Wireless had been asking questions about Racal, and in particular the Vodafone business, were confirmed by a friendly broker who'd spotted some atypical trading in Racal shares. He shared his suspicions with Harrison.

It appeared that Cable & Wireless hadn't forgotten how Racal had beaten it to the licence in 1983, and Vodafone's increasingly impressive performance just underlined what a lost opportunity that had been. Belatedly, it was looking to make amends. It took Cable & Wireless just two days to acquire a 2.8% stake in Racal – just below the 3% threshold, the point at which it would have been required by law to make its holding public.

By this time, Harrison was well aware of Cable & Wireless's intentions. He knew he had to raise the company's value significantly if it were to stand any chance of defeating a hostile bid, but the stock was already trading at a premium and, after years of missed targets, many investors remained sceptical.

Then Harrison remembered something an American investor had suggested to him a year earlier during a visit to the US. Racal should unlock the value within Vodafone by spinning it off as a separate company.

Harrison had dismissed the idea at the time, but now it seemed to be the obvious solution to the problem. There was no doubt that Vodafone, within Racal, was undervalued, suffering from a 'conglomerate discount'. Investors generally preferred pure plays – they were much easier to understand, for one thing – so an independent listing made sense.

Plans for a separate listing were put in place, and in April 1988 Racal announced that it intended to float 20% of Racal Telecom on the London and New York Stock Exchanges. This was no randomly chosen proportion – it was the minimum needed to avoid creating a tax liability for existing Racal shareholders and also permit the new stock to be included in the FTSE (Financial Times Stock Exchange) and MSCI (Morgan Stanley Capital International) indices, both of which track the performance of stocks. Racal's shares rocketed, gaining 66p that afternoon before climbing to a new high two days later.

But was it high enough to put it beyond the reach of Cable & Wireless?

Back in 1988, the business of valuing businesses wasn't as sophisticated as it is now, and the investment bankers brought in to handle the issue concluded they'd have to adopt the same approach as their US counterparts – and guess. To be fair, the discounted cash-flow calculations on which such guesses were based were quite well thought through, but there were still many within the UK's investment community who reacted with disbelief when it was announced that Racal Telecom was to be valued at £1.7 billion, 170p per share – way above the market average. This was actually equivalent to more or less $55 a pop – exactly where most publicly listed US cellular companies were trading at the time and somewhat below the $60 mark that McCaw had recently reached.

Looking back through the 1988 prospectus, it's surprising how conservative most of the company's forecasts were. The assumption was that most subscribers were going to be businessmen; there's little evidence to suggest that Vodafone's management expected there to be much of a consumer market at any stage in the near future. This kind of caution was somewhat at variance with the traditional Racal approach and it probably reflects Gerry Whent's business experience as much as anything else: he'd spent nearly 20 years at Racal selling expensive, high-margin stuff in small quantities to a small number of customers; the idea that something like Vodafone might ever become a mass-market consumer product was probably beyond his imagination. Well into the 1990s he remained far too cautious, suggesting that there might only be 5.5 million mobile-phone subscribers in the UK – just one in ten of the population.

His view wasn't shared by his younger colleagues, Chris Gent and Julian Horn-Smith, both of whom had a better understanding of consumer businesses and a far more expansive vision of the future, but as the financial community seemed to be struggling with the valuation of £1.7 billion, the more cautious voice prevailed.

The run-up to the listing wasn't without incident. A number of Racal's investors thought it unreasonable that they, as shareholders in Racal, should have to pay for something they thought they already owned, while others thought selling off just 20% was far too timid – the entire company should be spun off. Chief among these was Millicom, which had retained a small stake in Racal even though it had sold out of the Vodafone business. Nonetheless, Harrison and his colleagues felt that 20% was the right number.

A valuation of £1.7 billion might be enough to deter Cable & Wireless, but it was well within the reach of any of the regional Bells. A wholly independent Racal Telecom would have been easy pickings for any of the seven and a much more enticing prospect than a cellular company that was tied up with a number of struggling electronics businesses. Would the Bells have the imagination to make such a move? No one really knew – after all, they had only been independent for four years. It just wasn't worth the risk, so the company decided the original proposal should stand, Millicom was to be ignored and the whole initial public offering (IPO) circus got underway without further delay.

Investors in the UK remained lukewarm, but US fund managers took a different view. Many thought it a better bet than any of the American cellular stocks. As the roadshow moved across the country to the west coast, investors turned out in increasing numbers to meet the management and they liked what they saw. Only 20% of the 200 million shares on offer was available to US investors (for various legal reasons) but those 40 million shares could have been sold ten times over.

On 26 October 1988 trading in the new shares began. US investors went on to buy a majority of the 200 million shares that were available, giving Racal Telecom quite a high profile on Wall Street.

The threat from Cable & Wireless had been averted, at least for a little while.

CHAPTER 5

Consolidation in the USA

E ver since the first licences had been awarded in the USA in 1981, the regional cellular operators had been attempting to consolidate. Initially, this process had begun among the non-wireline operators, but by 1984 players from both sides of the industry were involved and a 'land grab' – literally – was getting underway. The regional nature of the licences – and also of the Bell companies themselves – meant that no one had a mobile footprint that came close to covering the whole country. Even though most calls were made close to home, the operators realised that a national presence would be highly desirable, albeit very hard to create.

One of the first to try to build such a national presence was John Kluge of Metromedia Inc. Always a contrarian by inclination, Kluge took issue with the prevailing opinion that the car-phone market had a 10-year payback period; he thought it was considerably shorter and, beginning in 1983, he set about investing some $300 million in the new industry, creating what at that time was the largest cellular business in the US. Perhaps the most notable deal of the several Kluge completed during this burst of activity was the November 1984 purchase of Graphic Scanning's franchises in Boston and Worcester in Massachusetts, for $48 million, equivalent to just $10 per head of population – somewhat on the low side.

Kluge soon had a rival for the top spot. During the course of 1979, a young entrepreneur who had already created a thriving cable-TV business had become interested when he heard that the FCC intended to award licences to operate cellular mobile-phone systems. He formed a small company called Northwest Mobile Telephone (later Interstate Mobilephone Company) with a view to bidding for licences in some of the larger markets in the Pacific Northwest, where he was based. The name of this man was Craig McCaw, arguably the most successful of all the cellular pioneers.

The FCC awarded the first A-block licences in the autumn of 1983. McCaw's Interstate won the two it wanted most – Portland, Oregon and Seattle, Washington (McCaw's hometown). He had also applied for licences in four other markets – San Francisco, San Jose, Denver and Kansas City – through his own company, McCaw Communications. He won those too.

It was at this point that the FCC changed the rules, deciding on the lottery approach. The randomness of the proposed methodology didn't appeal to McCaw in the least and other bidders were equally unhappy. McCaw contacted his rivals and suggested they try to coordinate their applications so the lottery wouldn't be needed. So began a period of frantic negotiation among the bidders, as they sought to trade bidding rights for one market with those for another. At the end of the process, only three of the 57 target markets actually had to enter the lottery, and McCaw emerged triumphant again, with the licences he wanted, in Austin, Oklahoma City, Wichita and Tacoma.

Winning licence bids was one thing, adding customers to the network quite another. It was a hard slog, not helped by the four-figure dollar cost of a typical car phone and the $100 monthly bill for use of the system. By 1986, McCaw only had 14,800 subscribers across his 12 markets, who together generated revenues of less than $300,000. The young man still had the firm conviction that mobile was the coming thing, however, and always seemed capable of raising cash when it was needed. In this he was helped by one of Wall Street's more adventurous bankers, the remarkable Michael Milken of Drexel, Burnham, Lambert.

Milken, 'the king of the junk bond', revolutionised the corporate debt market. The 'junk bond' concept was simple: put such a high coupon on a chunk of public-market debt that investors were tempted to overlook the risk associated with the instrument, dazzled by a 12, 13, 14% yield. It's a great way of getting a high-risk/high-growth business financed and

McCaw made good use of it. Over the next two years (1987 and 1988), Milken and Drexel raised over $2 billion to fund McCaw's ventures. And certainly those who bought McCaw bonds made, if not a fortune, then a very healthy return.

Craig McCaw's optimism about the future of the industry wasn't shared by all. During the course of 1986, MCI Communications – which had been awarded several key non-wireline licences – decided to give up on the new service and focus instead on its core long-distance operations, which were growing rapidly. McCaw agreed to MCI's offer price of $120 million (which included MCI Paging), which gave the company six further markets, including Denver, Pittsburgh, Sacramento and Salt Lake City, and minority interests in another five. The deal was announced on 3 July, after which McCaw promptly sold MCI Paging for $75 million, leaving the net cost of the deal at a minimal $45 million.

The acquisition of MCI Airsignal established McCaw Cellular Inc (another MCI!) as by far the largest of the non-wireline operators and it was well on its way to becoming the largest cellular operator in the country. It had a presence in many key markets, including San Francisco, San Jose, Sacramento, Seattle and Portland on the west coast, Miami and Tampa on the east coast, and St Louis, Denver, Kansas City, Minneapolis, Buffalo and Salt Lake City in other parts of the country.

McCaw made a point of creating clusters of cellular markets by acquiring other properties that were near to or contiguous with its existing licences. For example, having been awarded the licences in Portland, Oregon, Seattle and Tacoma, McCaw then bought out six more properties in Washington State, three in Oregon and one in Idaho to create a near-seamless network from the Canadian border down to the California state line. This approach reduced both capital and operating costs, but also gave the company a competitive advantage.

A document from 1990 that McCaw filed with the Securities and Exchange Commission (SEC) explained this. 'When markets are contiguous, the Company may offer uninterrupted service. Expanded service areas permit subscribers to receive incoming calls and make outgoing calls from anywhere within the region. Wider areas of service give the Company a competitive advantage in cluster markets where the Company faces differ-ent wireline competitors which are unable to offer uninterrupted service.'[31]

Noting what McCaw was up to, a couple of the Bell companies real-ised that they could make acquisitions both inside and outside their region, although until it had been tried, it wasn't clear how the FCC or the Department of Justice might react. In 1984 BellSouth took its first tentative steps down this path, bolstering its presence in one of its home states with the creation of new cellular joint ventures with other local-exchange opera-tors in Chattanooga and Memphis, Tennessee. In February 1986 BellSouth bought a 15% stake in the ambitiously named Mobile Communications Corporation of America, which had interests in cellular properties in sev-eral key markets outside the BellSouth footprint, including Los Angeles, Houston, Milwaukee, Indianapolis, Rochester, Honolulu, Gary/East Chicago and Bakersfield. The regulators seemed untroubled by the move, so in February 1988 BellSouth bought the remaining 85% of the business.

PacTel struck next, in early 1986, with the acquisition of Communications Industries. This had an interest in the non-wireline operator in San Francisco. Although PacTel had its headquarters in this beautiful city, it didn't have a licence to operate a mobile system there, as this had been granted to GTE. Communications Industries also held licences for San Jose, farther down the coast, and Atlanta and Dallas/Fort Worth. This business would eventually become a joint venture between PacTel (subse-quently AirTouch) and McCaw.

Southwestern Bell (SBC) jumped on the bandwagon in early 1987, acquiring some of Metromedia's portfolio of properties for $1.65 billion. The price equated to just under $145 per head of population, rather more than Kluge had paid for them a year or two earlier.

Later the same year, PacTel was back in action again, acquiring stakes in several large Midwest franchises clustered around the Great Lakes.

Later in 1987, McCaw Cellular went public, raising $270 million through an IPO and simultaneously selling $600 million of 12.95% debentures. The prospectus stated that the company was loss-making and unable to service

---

31 McCaw Class A Common Stock Prospectus, 27 March 1990.

its debt, and warned that this could continue to be the case for several years to come. The shares were 'speculative and involve[d] a high degree of risk'[32] but this failed to deter investors. The issue was oversubscribed and traded up from $21.75 to $24.50 on day one.

In January 1988 McCaw acquired the Miami Beach/Fort Lauderdale franchise from Post Cellular Communications for $245 million; in May it bought API Print Corporation's cellular licences and assets; and in December it closed the acquisitions of both Charisma Communications Corporation and Maxcell Telecom Plus.

On 19 January 1989 British Telecom surprised everyone when it announced that it had formed a partnership with McCaw, acquiring a 22% stake in the business for $1.2 billion. Most investors in the UK were bemused – that amount of money for a minority stake in an unprofitable business? It didn't make any sense. American legislation prevented any foreign entity from owning a business involved in wireless, so there was no possibility of eventual control, so wasn't this just a waste of time and money?[33] BT would eventually have the last laugh, as the investment proved to be well judged even if it didn't develop in the way that the company had hoped.

Bringing BT on board gave McCaw much-needed additional resources. As the company's presence in the US expanded, McCaw began developing plans to build on the clustering concept and offer a complete national roaming service, such that any of its subscribers in any of the company's many markets could use their phone anywhere in the country.

There was one major problem. To be taken seriously by the business community, the service would have to be available in New York and probably Los Angeles too, neither of which were McCaw markets. The wireline operators in those places (NYNEX and GTE) weren't going to let go of the jewel in their crowns, but there was an alternative. A company called LIN Broadcasting controlled the non-wireline licence in both New York and Los Angeles and also those in several other large areas, including

---

32 McCaw Cellular IPO Prospectus 1987.
33 The legislation dated back to the 1930s and was designed to prevent foreign organisations broadcasting communist propaganda to the unsuspecting inhabitants of the home of the free.

Philadelphia, Dallas and Houston.

McCaw approached LIN and suggested a merger. The offer was turned down. So in late June 1989 McCaw Cellular announced a hostile bid: a $120 per-share cash offer for LIN, valuing the business at $5.85 billion. On Saturday 1 July LIN's management announced that it rejected the bid on the grounds that the valuation was inadequate. This was the beginning of the first great takeover battle in the industry's history.

In the following weeks, LIN was approached by BellSouth, one of the giants of the US telecoms industry. It proposed a 'white knight' deal to LIN, with the suggestion that the two companies' cellular interests be merged to create a new, jointly owned company that would immediately become the largest mobile operator in the US, with about 500,000 customers – some way ahead of McCaw's 366,000. It would have a presence in more than 40 of the largest metropolitan markets and would be in a position to offer the complete national roaming service McCaw had planned.

On 11 September LIN's management broke the news of the deal to Wall Street, announcing that ahead of the merger it would spin off its remaining broadcasting assets to its shareholders and also pay them a one-off dividend of $20 a share.

McCaw suddenly found itself up against one of the largest companies in the country. It paused for a moment, then on 20 November raised its bid, offering $150 per share for 22 million shares. These, together with the 9.4% stake it already owned, would give McCaw an approximate 52% controlling stake in the company. The offer was not entirely straightforward as it included (inter alia) a provision that allowed the remaining shareholders to sell their stakes in LIN to McCaw in five years' time, at a price to be independently determined by reference to then-prevailing open-market valuations.

BellSouth threw in its hand, terminating its offer on 11 December. The president of BellSouth Enterprises, the subsidiary company that ran the corporation's unregulated activities, commented somewhat ruefully, 'Our merger would have created the premier cellular company in the world, but such decisions ultimately must reflect economic and strategic realities.'[34]

The acquisition of the controlling interest in LIN Broadcasting made McCaw the clear number one in the new mobile market, with an enlarged business a footprint that contained over 100 million potential customers. McCaw had won but was saddled with debts of more than $5 billion, and now that the dust was settling, Wall Street didn't like what it saw. The company tried to calm nerves through the disposal of a few of the less important licences, raising $1.3 billion, but its share price came under severe pressure, plummeting from $36 to $11.50 in just a few weeks, leaving the equity with a value of just $2 billion, down from $6 billion at the time of BT's investment.

While McCaw struggled to digest its new acquisition, life went on around it. In July 1990 GTE bought Continental Telephone, propelling it into second place in the market, behind McCaw. And in late 1991 Bell Atlantic announced the acquisition of Metro Mobile CTS, the second-largest independent cellular business in the US, in a deal valued at just under $2.5 billion. The deal gave Bell Atlantic a further 180,000 cellular subscribers to add to its existing 283,000.

This transaction, which closed in April 1992, brought an end to the American mobile industry's first phase of consolidation. At the time these seemed like big deals, but they would be dwarfed in just a few years by a series of moves that would transform the telecom landscape; not just in America, but beyond.

---

34 *The New York Times*, 12 December 1989.

CHAPTER 6

The transition to digital

The switch from analogue to digital was a revolutionary transformation of an entire industry. Digital systems offer higher-quality sound. They make better use of the available radio spectrum and hence are able to support a larger user base. Digital components offer the possibility of greater integration at the chip level, leading to lower power consumption, so battery life is better. Digital systems are inherently more secure and less susceptible to cloning or eavesdropping, while they can also support a greater range of services.

An international digital standard would have the advantage of international compatibility. Because standard devices would be produced in greater volumes than any single proprietary technology, scale efficiencies could be expected, leading in the long run to lower unit costs.

Today's mobile handsets have fulfilled this promise. They're small and elegant, but they're only that way because all the circuitry has been squashed into a single dedicated microcircuit. The functionality of that circuitry rendered in discrete components would produce a unit about the size and weight of a chest of drawers that dims the lights when you turn it on.

From the birth of the first European mobile networks in the 1980s, the industry was on the search for a digital standard that would solve the problem of incompatibility between networks in neighbouring countries. The European Conference of Postal and Telecommunications Administrations (CEPT) had created a study group, Groupe Speciale Mobile (GSM or Special Mobile Group), in 1982 to work towards a common European standard based on digital technology. The leading phone companies of Europe nominated representatives to the GSM.[35]

By 1987 the GSM had agreed to focus on a technique called time-division multiple access, or TDMA, one of several possible methods of multiplexing separate signals. The technology is complex, but the concept is easily explained by the following analogy. The transmission of multiple signals is like a room where numerous people want to talk to each other at the same time. One way to make this possible is for individual speakers to be disciplined and only talk in turn (time division); another option is to use different pitches (frequency division); while a third, more radical solution, is to speak in different languages (code division). Those in the room who understand the language will listen, the rest will tune out.

By 1989, oversight of the project had passed to a body called the European Technology Standards Institute (ETSI), and by 1991 the first networks based on the new GSM standard had started operating.

That the various European administrations had managed to collaborate to create a common standard was a minor miracle, given the numerous vested interests involved: the operators, the bureaucrats, the various supervisors and policy-makers, and the manufacturing companies, most with their own proprietary technologies. These included Siemens from Germany, Sweden's Ericsson and Alcatel from France; the UK was represented by GEC Plessey Telecommunications (GPT) and Standard Telephones and Cables (STC) (which in 1991 was to become part of the Canadian giant Nortel). American interests were represented by AT&T, which had re-entered the European market in a partnership with Dutch electronics firm Philips; and Motorola, the leading handset manufacturer. Finally, there was an obscure Finnish company, best known for its tyres and forestry products, but which had a small sideline in consumer electronics, called Nokia.

One key element of the new GSM standard was that the identity of the user wasn't defined by the handset itself, but rather by a 'subscriber identity module' or SIM. This allowed subscribers to use more than one device,

---

35 The meaning of the GSM abbreviation was later revised to 'Global System for Mobile' once it became clear that that wasn't too much of an exaggeration.

switching the SIM from, for example, a hand-portable phone to a car phone or a data modem. Later it would enable customers of one network to leave and join another without the need to buy a new handset.

Despite the standard's many advantages, initial uptake was slow. There were several reasons for this. The new GSM handsets were much more expensive than the latest analogue designs, and a lot larger – Motorola had set the bar high when it released its diminutive (350-gram) MicroTAC handset in 1989, and all the new digital devices failed to get close to it.

Moreover, coverage didn't match that of the existing cellular systems – in more than a few networks, the 'Swiss cheese' effect ('holes' in the network) disrupted the best efforts of the network planning teams. Many network operators thought that they could simply locate a digital transmitter on the same site as the existing analogue transmitter and automatically achieve the same coverage. Not so. Analogue signals faded gracefully over distance and could still be picked up some way beyond their specified coverage range, whereas digital ones just stopped dead. It was either there or it wasn't. This left areas at the edge of coverage where only the analogue signal could be received, creating holes in the network.

And there was another problem: the GSM solution wasn't alone in the market. Digital AMPS (D-AMPS) was an American development, a fairly straightforward adaptation of the US AMPS standard using the existing AMPS channels but dividing these into three separate time slots (time division) and then compressing the voice signal to achieve a threefold increase in capacity, combined with greater security. Compatibility with existing AMPS networks added to the attraction of what was a rather lacklustre technical solution.

Personal Digital Cellular (PDC) was similar to both GSM and D-AMPS, in that it was also a TDMA-based standard, but the Japanese developers had put the need for very small handsets above the needs of the network as a whole. The result was very attractive phones that didn't work very well unless the radio signal was very strong. Although the only country where PDC was ever deployed was Japan, at its peak in September 2003 it had over 62 million users, close to the 74 million total AMPS managed across all 98 countries where it was in service.

The third competing technology, code-division multiple access (CDMA), divided the industry into two opposing camps, essentially the US and the rest of the world (although some in the US were firmly in the TDMA camp); PacTel was the most committed of the CDMA supporters but it was by no means alone. Outside the US, only SK Telecom in South Korea had committed itself to CDMA by the end of 1993.

CDMA is a complex technology based on 'spread spectrum', where a signal is spread equally over the entire available spectrum. To prevent interference, each signal is given a unique code, thus allowing multiple signals to be transmitted simultaneously over the same slice of spectrum. CDMA's proponents maintained that it was superior to both GSM and D-AMPS, as it offered the possibility of a twentyfold increase in network capacity compared to the AMPS standard it hoped to replace, as against the three-to-one advantage offered by D-AMPS, or GSM's (theoretical) six to one. This and other claims were justified, but there was a further problem – equipment based on the new standard wasn't quite fully developed and certainly wasn't ready to go into production.

Europe was committed to GSM, while Japan was going down the PDC route. Some operators in the US had concluded that they couldn't wait for CDMA, as they needed the advantages of a digital system straight away. McCaw was very much in that camp, and as early as September 1990 it placed a huge order with Ericsson for equipment to facilitate the transition from AMPS to D-AMPS. A lot of the rest of the industry was still undecided.

By the time the first GSM networks in Europe had been running for a few months, many operators in other parts of the world had seen enough to decide in favour of the European standard. Telstra from Australia and SmarTone, a new operator in Hong Kong, became the first outside Europe to launch at the beginning of 1993, well before most European operators. New Zealand followed suit, launching a new network based on the European standard. That company later became part of Vodafone, but in 1993, it was called BellSouth New Zealand, making this the first occasion when an American company selected a technology not developed in the USA. The giant Bell company would later make the same choice in its home market.

The following year there were GSM launches in Iran, Morocco, Pakistan, the Philippines, Qatar, Russia, Singapore, South Africa, Vietnam, Indonesia, Thailand and the People's Republic of China. GSM arrived in India in 1995, by which stage the European technology had been adopted as the standard by countries with comfortably more than 50% of the world's population.

Technology choices might have obsessed the network planners within the world's mobile-phone companies, but the average consumer could hardly care. The advantages of the new digital phones weren't all that obvious and most people were happy to stick with their analogue devices, not least because they were smaller and lighter – and had cost a lot to buy in the first place. By the end of 1993, digital systems had just over 1.6 million users worldwide, out of a total of 34 million – about 6.5% of the market. A year later the picture was better, but not by much – the number of digital phones had roughly quadrupled to 6 million, taking its share of the total to 13.6%. Looked at another way, of the 20.8 million new customers that year worldwide, 16 million had plumped for an analogue handset.

The year 1995 was better for the digital camp. The number of digital devices nearly tripled to 17.6 million, nearly a quarter of the 87-million-strong market; digital accounted for more than a third of all new connections. And by the end of 1995 fully two thirds of all new customers had gone digital, raising the overall base to just under 50 million, or 37% of the 136-million total.

A simple explanation for this is that there were now more digital networks in place, but there's also a more subtle explanation for what was happening, which points to a sea change in the way mobile phones were perceived. When Nokia launched its iconic 2110 handset – a thing of beauty when compared to anything else on the market, with an elegant design that still looks good today – almost everyone who saw one wanted one.

This began a trend that's lasted to this day, where the appearance and specification of a mobile handset are more important than its qualities as a communications device; mobiles, especially higher-end smartphones, are a fashion item rather than a business tool, and handset manufacturers ruthlessly exploit this fact, with marginal tweaks to the appearance and performance designed to maintain a steady replacement market.

CHAPTER 7

Vodafone's path to independence

Although the UK's mobile-phone duopoly was providing a reasonable service at a reasonable price, the government was keen to introduce more competition into the market.

It had already taken a step in this direction in July 1988, when it announced that it planned to offer four licences for a new service called telepoint. Using the European CT2 digital cordless-phone standard (then under development), a telepoint phone would allow subscribers to make calls if they were within 100 metres of a base station – a telepoint. The system wasn't able to hand off calls from one telepoint to the next and couldn't handle incoming calls, but these limitations were offset by the small size and low cost of the handset (around £250) and the lower monthly subscription fee. It was intended to be a consumer product, the equivalent of 'having a phone box in your pocket'.[36]

Eleven bidders came forward and four were chosen. Ferranti won a licence for its Zonephone proposal, while the other three went to consortia, all of which were backed by industry heavyweights – Philips had teamed with Barclays Bank and Shell in BYPS Communications;[37] STC led Phonepoint, a team that included BT, France Telecom and NYNEX; and

---

36 Attributed to Lord Young, Secretary of State for Trade and Industry, July 1988. Young later went on to chair Cable & Wireless.

37 Hutchison Whampoa tried, and failed, to make a success of one of the telepoint operators, buying BYPS and rebranding it 'Rabbit'.

Callpoint consisted of Motorola, Mercury Communications and the rather less celebrated Shaye Communications.

Although the few who used the service liked it, and especially liked the digital cordless phones, which were a step up from anything available at the time, telepoint was a complete commercial failure. However, it introduced several new players in the UK telecoms market, some of which went on to feature in the next phase of deregulation.

In late 1989 the UK's Department of Trade and Industry stated that it intended to increase competition in the mobile market through the offer of no fewer than three new licences. According to the department's document *Phones on the Move*, the new entrants wouldn't be operating 'old-fashioned' cellular-telephone systems, but personal communications networks (PCNs), using an exciting new technology called digital communication system 1800 (DCS-1800).[38]

The use of the terms PCN and DCS-1800 was really quite unhelpful: the existing cellular systems were also personal communications networks, and DCS was merely the GSM standard but at twice the frequency. Nonetheless, as *Phones on the Move* noted, 'The UK has more than 20 million telephone subscribers who access the public telephone network through a copper wire link. There is no inherent technical reason why these subscribers should not access public telephone networks through mobile radio means.'[39]

Twenty million! No one had spelt it out quite like that before. And there being 'no technical reason' flew in the face of conventional wisdom that claimed there would never be sufficient radio spectrum to support a mass market.

The licences went to three teams, each of which had the backing of a large telecom operator: Mercury PCN, a consortium led by Cable & Wireless, which included Motorola and Telefónica from Spain; Unitel, headed by US West and including Deutsche Telekom, STC and THORN EMI, the large

---

38 The DCS standard was later referred to as GSM-1800. The number refers to the operating frequency.

39 *Phones on the Move*, a consultation document produced by the British Department of Trade and Industry DTI in 1989.

UK entertainment conglomerate; and Microtel, which had been formed by the Space and Communications division of British Aerospace (now BAE Systems) and included PacTel, Millicom and the French defence contractor Matra SA.

After the initial excitement of the licence award, the consortia members began to realise the difficulties they faced breaking into a market with an untried – and as yet unavailable – technology. Telefónica was the first to abandon ship, dropping out of Mercury PCN. Deutsche Telekom was next, leaving Unitel. Then PacTel, Millicom and Matra all announced their departure from Microtel, leaving British Aerospace as the sole shareholder. These were soon replaced, however, with a single new partner, Hutchison Telecom, a subsidiary of the Hong Kong conglomerate Hutchison Whampoa, which owned one of the three mobile businesses in its home market but harboured ambitions to be a player on the world stage. It took control of Microtel, though British Aerospace remained a junior partner for a little while longer.

Meanwhile, Mercury PCN and Unitel were considering their options. Although the two had very different views of how to address the market, they had very similar needs, and eventually this would drive them into each other's arms. A merger was announced in June 1992 – probably not coincidentally, the very month that Vodafone launched its first commercial GSM service in the UK. Three of the five remaining shareholders (Motorola, STC and THORN EMI) took this as an opportunity to slip away from the party, leaving US West and Cable & Wireless as equal partners in the venture, now named Mercury one2one.

In the three years following the IPO of Racal Telecom, the Vodafone business continued to make good progress, bettering the expectations of even the most optimistic commentators. It passed its original target of half a million customers in April 1990, well ahead of schedule. It continued to run rings round Cellnet, in both market share and profitability: it would announce annual profits of £165 million for its 1989/90 financial year, while its competitor was still barely profitable.

Yet once again its share price had begun to languish. The state of the UK economy wasn't great and the news that the government intended to award three new mobile network licences didn't help. Subscriber growth slowed throughout 1990 and several analysts took this as a sign that they'd been right all along – BT's revised prediction that there would never be more than 3–4 million mobile subscribers in the UK began to be repeated by people who thought themselves experts.

Racal's management knew how short-sighted these assertions were, but they couldn't do anything about the markets, even though they had a keen sense of the very real danger the lacklustre share price represented. They suspected they might have to deal with another predator, as they had three years earlier. They didn't have to wait long for their fears to be realised.

In the early summer of 1991, although it had its own mobile business by now, Cable & Wireless came back again. This time it had brought some extra muscle in the shape of GEC, Britain's largest electronics company. Arnold Weinstock, GEC's long-serving managing director, had a prodigious memory and hadn't forgotten that Racal had bested it in the 1980 battle for control of the Decca Company. He wasn't interested in the phone business – Cable & Wireless could have that; he wanted revenge and his target was the defence assets. The two made a formidable team and represented a very clear threat to Racal.

But Racal could move incredibly quickly when it needed to. Harrison immediately summonsed the board for an emergency meeting on Sunday 11 June. He suggested to his colleagues that the Racal Telecom business was now big enough to stand on its own two feet and that, were it to be cut loose, the danger of a hostile takeover of both Racal companies – Racal Electronics and Racal Telecom – could be averted.

The next day plans for a complete demerger of Racal Telecom were posted with the London Stock Exchange.

Chris Gent knew nothing about this until later that morning, when an analyst called to congratulate him on his imminent independence. Somewhat taken aback, he called Gerry Whent, who was on his way to Australia to promote Vodafone's bid for the third mobile licence in that country. Whent picked up the voicemail when he landed in Singapore and returned the call. It took only a moment for Gent to realise that his boss knew nothing about the plan either – even though he was a main board director of Racal!

But Harrison was right – the proposed demerger had the desired effect and the rising prices of both Racal stocks put the business out of reach. Once again a predator's threat had been avoided.

Racal Electronics distributed its 800 million shares in Racal Telecom to its shareholders in September 1991 and the former subsidiary formally adopted a new name, Vodafone Group plc. It immediately became a constituent of the London Stock Exchange's FTSE-100 index.

In September 1993 the Mercury one2one network eventually went live, but to the surprise of many, the service was available only in the area inside the M25, the orbital motorway that circles London. The explanation was that the DCS-1800 network technology required many more base stations than GSM to achieve national coverage, so the company had chosen to target the largest market first and then roll out the service across the rest of the country. According to one2one, the geographic limitation wasn't that important, as most calls from mobiles take place within a few miles of home, and the area inside the road accounted for over one third of all UK calls.

And there was a further surprise. Those who subscribed to the 'PersonalCall' tariff would be treated to free off-peak calls for a limited period. That was revolutionary – free calls! At the press launch, a journalist asked how long the offer would last. Snatching the microphone from one of the company's representatives, Lord Young, the chairman of Cable & Wireless, went way 'off message' when he told his incredulous audience – and equally flabbergasted colleagues – that it would last forever. Subsequent promotional material summed this up rather dryly: 'This offer expires after you do.'

But the public didn't buy the limited-coverage argument: people wanted a phone that would work anywhere. The free-call offer was a nightmare too – it was costly to operate and attracted all the wrong kind of customers, people who only used the phone to make the free calls. There was even talk that some people had bought two of the things and were using them as remote baby alarms. Over the years, Cable & Wireless was to spend a small fortune buying subscribers out of these early contracts.

In the meantime, Hutchison had taken its time building its network to ensure that it could offer something close to national coverage when it eventually launched. This it finally did, in April 1994, Microtel becoming Orange and launching the world's first national DCS(GSM)-1800 network.

The launch was supported by a powerful and really rather weird advertising campaign, with a voice repeating, 'One, two, three, four,' over and over again to repetitive music, as various apparently random images flashed across the screen. But if the advert borrowed from modern minimalism, the rest of the package was entirely new, and there was much about it to like. An Orange subscription came with a bundle of 'free' minutes (they weren't actually free because a user couldn't take the minutes on their own without a subscription), and the service included voicemail, call-waiting, caller identification and many other new features. Importantly, calls were billed by the second, not rounded up to the next full minute. Subscribers could choose an 1800MHz variant of Nokia's new 2110 handset.

Orange's biggest achievement was that it managed to persuade people that mobile phones weren't just for the businessman or the more fortunate – they were for everyone.

One of the effects of the new UK licences was that the USA began reconsidering whether two licences were really enough. It decided they were not.

In October 1993, a decade after the first cellular launch in the US, the FCC announced that it intended to increase competition in the market by auctioning additional licences. Like the UK licences, these weren't for plain old cellular, they were for an entirely new service – the so-called personal communication service (PCS). 'PCS relies on a network of transceivers that may be placed throughout a neighborhood, business complex or community to provide customers with mobile voice and data communications,' PacTel explained to its shareholders in its 1993 Form 10-K (a more formal version of an annual report to shareholders, submitted to the SEC).

The FCC initially established two different sizes of service areas nationwide for PCS: 47 large areas referred to as major trading areas (MTAs) and 487 smaller ones, each a subdivision of an MTA, to be known as basic trading areas (BTAs). When the auctions eventually began, however, it turned out that there were 51 MTAs and 493 BTAs. The spectrum reserved for the new service was in the 1.9GHz band and had been divided into no fewer than six separate blocks, designated A to F (so in any one market there might be as many as eight separate operators – six new PCS companies plus the existing two cellular companies). The A and B blocks were allocated to operators in the 51 MTAs, while the C, D, E and F blocks were allotted to the smaller BTA regions. Thus, in addition to the original patchwork quilt of 1,468 separate cellular licences, there were to be 2,074 more.

There were three further stipulations that made these auctions problematic. First, the A, B and C blocks contained three times more spectrum than the D, E and F blocks, with 30MHz of available bandwidth against just 10MHz.

Second, the C and F blocks weren't available to existing operators but were reserved for 'designated entities' – companies owned by minority groups. To ensure that the incumbent operators didn't squeeze new entrants out of the market, the FCC limited the amount of PCS spectrum any existing cellular operators could acquire to just 10MHz. This made sense, but as the new market divisions weren't the same as the earlier CMA arrangement, assessing whether any one operator had breached the spectrum cap in a specific service area wasn't always an easy task.

Third, and most problematic of all, there was no obligation to use any particular technology: each operator could decide which of several competing standards they preferred. This meant that customers might find that they needed two or more phones to guarantee continuous service as they travelled from place to place. As Western Wireless put it in its 1996 10-K, 'A subscriber of a system that utilizes GSM technology is currently unable to use a GSM handset when traveling in an area not served by GSM-based PCS operators, unless the subscriber carries a dual-mode handset that permits the subscriber to use the analog cellular system in that area. Such dual-mode handsets are not yet commercially available and will be more expensive than single-mode handsets. [40]

None of this might have mattered much had the mobile market looked like it does today, with the US accounting for only 5–6% of the mobile phones on the planet. But back then, the 24 million phones in service in America constituted well over 40% of the global total, so what the Americans did mattered a lot.

Some of the harsher critics of that FCC decision suggest that what the Americans did created chaos, putting back the advent of a genuinely global phone for a generation – but everything was still very new then, and in the mid-1990s it wasn't at all obvious that the GSM standard would become such a dominant force in the market. Others suggest that the blame for the failure to achieve a 'global system for mobile' back in the 1990s lies with the International Telecommunications Union (ITU), founded in 1865 to facilitate international connectivity in communications networks (and now a part of the United Nations), as it failed to harmonise standards and establish coordinated frequency blocks. One way or another, however, the arrival of digital technology didn't solve the industry's incompatibility issues.

For the operator community, the auctions were something of a mixed blessing. On the one hand, they provided an opportunity to acquire licences in new markets. On the other, they represented both additional competition and additional, unwanted licence costs. Moreover, as most of the major players in the market wanted to achieve national coverage, competition for the most valuable markets would probably be intense.

To reduce the number of possible competitive bidders and also to improve their chances of achieving national coverage, four of the largest cellular companies agreed to pool their resources. AirTouch joined up with Bell Atlantic, NYNEX and US West to form a consortium called Prime PCS. Sprint adopted a similar approach, creating Wireless Co. out of a new Sprint subdivision, Sprint PCS, together with three large cable companies Tele-Communications Inc (TCI), Comcast and Cox. This strategy

---

40 Western Wireless 10-K filed with the SEC on 31 March 1997.

clearly worked, and the two groups limited their outgoings to $1.11 billion and $2.11 billion respectively.

The PCS auctions were to have another adverse effect – the industry's first bankruptcies. A number of new companies entered the auction with inadequate capital, in the hope of making a quick return by reselling the licences before they were required to pay the full bid price. NextWave Wireless was typical: it won licences for numerous major markets, including New York, Los Angeles, Boston, Houston and Washington, at a total cost of over $5.6 billion. It managed to pay the minimum 10% down payment, but then defaulted on the rest and filed for bankruptcy. The FCC confiscated the licences and resold them. (What wasn't typical in this case is that, remarkably, NextWave took the FCC to court, claiming that the confiscation had been illegal, and at the end of a convoluted and extended legal battle emerged triumphant. It reclaimed the licences and promptly sold them back to the operators that had acquired them from the FCC a decade earlier – but at a very much higher price.)

Although the auction generated over $23 billion in fees for the US Treasury, it can hardly be considered a great success.

CHAPTER 8

Competition encourages growth

Towards the end of the 1980s, the governments of Europe agreed to introduce competition into their mobile markets. The launch of the new GSM networks provided an ideal opportunity for this. Not all administrations were happy with the decision, but the difference between the market in the UK and elsewhere in Europe was just too marked to be ignored. Mobile was already a vibrant market in the UK; elsewhere in Europe, it was close to moribund. To quantify – in December 1987, after three years of mobile service, only 0.063% of the German population owned a mobile phone. It was half that level in Italy (0.029%), although France had a marginally more respectable figure of 0.073%.

In the UK, by contrast, penetration was 0.44%. In absolute terms, that equated to more than 250,000 telephones – comfortably more than twice the combined totals of the other three big European markets (104,000). And the UK numbers didn't suffer in comparison with those of the US; indeed, they were better. It had taken the US almost two years longer to reach the same level as the UK, which it did in the third quarter of 1988. Competition, it seemed, worked, so perhaps telecommunications wasn't a 'natural monopoly' after all.

During the first half of the 1990s, administrations in all the major western European countries took turns to announce plans for a second mobile licence. Vodafone and its American rival PacTel were probably the only two companies that appreciated exactly how valuable these opportunities would prove to be. The months that followed each of these announcements were characterised by frantic negotiations as would-be bidders struggled to form consortia to apply for the new licences. When building a team, a strong local partner was all-important – that partner wasn't going to have any telecom credentials, of course, as the local markets were monopolies, so prospective bidders would look to choose a company with good political, economic or financial connections.

France had been one of the first markets to endorse competition, when in 1988 it awarded a second licence to a consortium headed by the Compagnie Generale des Eaux (CGE, later Vivendi) to operate an NMT-450 network. CGE had taken the licence on the understanding that it would subsequently be awarded a licence to operate the new GSM system when the digital technology had been finalised.

The original CGE consortium, the Société Francaise de Radiotéléphones (SFR), benefited from a typically convoluted shareholder structure. CGE held its stake through a joint venture with Southwestern Bell, while other investors included Alcatel, the French bank Credit Lyonnaise, a Belgian cable-TV operator called Coditel, the state-owned Télédiffusion de France (TDF), BellSouth International, and Vodafone.

The British company was only allocated 4% of the business initially, but in early 1989 another two opportunities arose for new networks, in Greece and in Malta. Vodafone signed memoranda of understanding with the governments of both states in preparation, only to find that the Greeks had changed their minds. Malta went ahead as planned, however, with Vodafone entering into an 80:20 partnership with Telemalta, the local phone company. It wasn't a huge opportunity, even allowing for the 80%, but it had a significance over and above the market opportunity: this was the first time that Vodafone had been selected on merit, beating all comers.

Gerry Whent's decision to entrust the job of international expansion to Julian Horn-Smith was looking increasingly astute, and he set the company a new target, that of creating as large a presence overseas as it had in the UK.

One of the first and biggest opportunities for Vodafone arose in Germany, which was an undeveloped market, with just 99,900 mobile connections at the end of 1988 (a penetration rate of just 0.13%). (The UK, by contrast, had nearly five times that many and a penetration of 0.9%.) Various consortia were created and seven bids submitted. As the year wore on, it became clear that two were some way ahead of the pack: BMW, which had brought in Veba AG, a German electrical utility, and Vodafone, to provide the industry expertise; and Mannesmann, with PacTel and a local bank. In an attempt to neutralise British lobbying, it had also co-opted Cable & Wireless, which had taken a token 5% stake.

The Mannesmann-PacTel team was awarded the licence, thanks in large part to Jan Neels, the president of PacTel International. Astute and obviously well connected, Belgium-born Neels believed that the government wanted to give a shot in the arm to the economy in the north: Mannesmann was based in the northwestern city of Dusseldorf, while BMW was located in the prosperous south.

That a key lobbyist for the BMW team had died in the weeks leading up to the decision did not help. It was the first but not the last time that Vodafone would lose a key licence to PacTel.

While Europe represented the largest opportunity for Vodafone, the idea of an open market for telecom services was also beginning to be accepted in other parts of the world. Some observers saw this next phase of the market's development as being akin to a giant game of Monopoly, where players went around the world trying to acquire the best properties before their competitors snapped them up. But it's not easy to win the game if you don't have a coherent strategy: not all properties on the board offer the same return, and the cost of doing business and barriers to entry vary considerably.

Obviously, there was a preference for the wealthiest countries where citizens were most able to afford the new service.[41] But with fewer natural advantages than its many competitors in terms of its size, brand, existing

---

41 Later, after the opportunities in the top-tier markets had all been taken, Vodafone put more emphasis on those markets with the most growth potential, especially those that weren't well served in traditional telephony by an incumbent monopoly.

infrastructure, personal connections, and government backing, Vodafone was well aware that it had to choose its targets carefully. Almost all its potential competitors were far larger than it, had long-established telecommunications businesses with substantial asset bases, had better links to the rest of the industry, and had all of the benefits that government ownership bestows.

In late 1991 Vodafone was approached by its old partner Millicom, offering the company a 30% stake in Pacific Link in Hong Kong for $75 million, or $40 per pop. The business was the newest and smallest of the territory's three mobile operators, but it was growing rapidly. At the time Vodafone bought in, there were 42,000 subscribers on its D-AMPS network. By the end of financial year just over four months later in March 1982, this had risen to 58,000.

At around the same time, the licensing process in Greece that had been deferred in 1989 came back to life. The original monopoly proposal had changed to a duopoly and analogue had changed to digital, but it was still an attractive opportunity. Vodafone submitted its bid in early 1992 and was successful. It won one of the two available licences, taking a 45% stake in the company that would become known as Panafon. Its team included France Telecom with 35%, and two Greek businesses, Intracom and Data Bank of Greece, a local IT company, which shared the remaining 20%. This was to prove a great success over the next few years.

In 1992 the government of India had decided to award licences for regional GSM systems, starting with the four largest metropolitan areas, Mumbai, Kolkata, Delhi and Chennai. These licences hadn't really been on Vodafone's radar, but RPG Enterprises, the Indian conglomerate that headed the bid, was persuasive. Vodafone agreed to take a 20% stake in a new company, RPG Cellular, which would go on to win one of two licences in Chennai.

Then, in the same year, a second opportunity arose in Germany. The third digital network (after Deutsche Telekom's D1 and the new operator, Mannesmann's Mobilfunk) would use the same DCS-1800/GSM-1800 technology employed by the PCN operators in the UK.[42] A number of the companies that had been unsuccessful in the 1990 contest teamed up to bid for the new licence. BMW was one, and this time it recruited Hutchison and GTE; and BellSouth teamed with Thyssen, a heavy-engineering company based in Dusseldorf.

Julian Horn-Smith thought that he was in danger of being sidelined as the teams came together, so he roped in Veba, the energy company that had been part of BMW's original consortium and, not to put too fine a point on it, the two edged their way into the BellSouth E-Plus consortium. This duly got the nod.

Vodafone's profile was improving all the time and its track record in international pitches was looking ever more impressive. On those occasions when it found itself on the losing side, the victorious consortium invariably included PacTel.

Vodafone had been particularly unlucky in Italy. In 1993, under the name Unitel, it had put together a strong team including the giant car company Fiat, and Fininvest, an investment company controlled by media tycoon Silvio Berlusconi. It faced competition from four other groups. Olivetti, the Italian typewriter-to-technology conglomerate, had joined forces with Bell Atlantic to form the Omnitel consortium, and that looked to be the biggest threat. Naturally, the Californians were there too, though by this time they had swapped the PacTel name for the rather more whimsical AirTouch, following their demerger from Pacific Telesis a few months earlier. AirTouch's Pronto Italia team included the Banco di Roma, a long list of Italian corporates, including Benetton, and a somewhat reluctant Mannesmann.

Unitel was cruising towards victory but then Berlusconi announced that he was entering politics, heading a new political party called Forza Italia. To the amazement of almost all the pundits, he won a snap general election – on the same day the licence was due to be awarded. To avoid the obvious

---

42 It costs roughly four times more to cover an area with GSM at 1,800MHz than it does at 900MHz, but once coverage has been achieved, there's greater capacity on the network. For Vodafone, a third licence in a big market such as Germany was still worth having, even if the second would have been preferable.

charge of nepotism, it had to go to the Omnitel team.

This loss was particularly irksome as Vodafone didn't feel that it had been beaten by a better team. AirTouch had far greater political reach, but was it any better at the mobile-phone business? Not really. But in some ways these occasional losses were inevitable. AirTouch could – and did – call on the US government to support its efforts overseas, while Vodafone was pretty much on its own. If the UK government deigned to lend a hand, it would help the company it saw as a national champion. This was some-times BT but more often Cable & Wireless, never Vodafone. Ironically, as Cable & Wireless was always in a different team to Vodafone, the best efforts of the government usually hindered, rather than helped, its most successful operator.

The shareholdings Vodafone had accumulated at this point – France, Malta, Hong Kong, Greece, India and Germany – put it about a third of the way towards its target of matching its overseas market to its home one. Then, suddenly, one after another, four new licences presented themselves in quick succession.

In Australia, the state-owned monopoly held the first licence, while a team headed by Cable & Wireless and including BellSouth had bought Australia's international satellite company and was given the second licence as part of the deal. Vodafone's 95%-owned Australian subsidiary bought the third licence for A$160 million.[43]

In the same part of the Pacific Rim, BellSouth was to beat Vodafone to take the second licence in New Zealand, but Vodafone took the honours in Fiji, where it acquired a 49% stake in a GSM monopoly.

The fourth opportunity was in South Africa, where, in 1993, ahead of the election of the new democratic government, representatives of Nelson Mandela's party decided to open up the telecoms market by awarding two

---

43 Acquiring a majority stake in an international business was, at this early stage, rather unusual. Most of the time, Vodafone and its competitors had to content themselves with a minority shareholding. In practice, this didn't present any real difficulty, as the operators developed a technique called 'minority management' – they would allow their partners to take all the main management roles except network design and marketing, and, provided the network was built properly and the service marketed properly, everything else fell into place.

new GSM licences. The incoming South African government wanted an international partner to help Telkom, the national PTT, operate the new GSM system, and in July 1993 Vodafone was selected from the shortlist. A new company, Vodacom, was formed, in which Vodafone took a 35% interest. (The second licence went to Mobile Telephone Networks Holdings, or MTN, a consortium that included Cable & Wireless.)

Vodafone was now very close to its target as far as its international footprint was concerned. Acquiring this many international properties was a huge achievement, given its limited resources when set against those of BT, France Telecom and its numerous American competitors. The company had taken risks on an almost daily basis, without any kind of safety net.

When the results for 1994/95 were published, however, they weren't bad but it was clear that they could have been better. As the various new international businesses were at an early stage of development – only three had actually launched – they were incurring start-up losses and, taken together, these amounted to £45 million, or about 15% of the group's £340-million total pre-tax number.

There was a rather nastier problem, though. A subtle but massive fraud had hit Vodafone's in-house service provider Vodac, with unscrupulous dealers 'connecting' nonexistent customers to claim higher bonuses. This put a £40-million hole in the company's accounts and obliterated its profits. Annoyed investors responded by calling on the company's management to focus on the UK and spend less time on its international adventures.

Vodafone took heed of the advice, because it knew the fraud was merely a symptom of a deeper problem that needed to be addressed: malpractice of this nature probably would have been easier to spot in a smaller, simpler organisation, such as Vodafone had been a few years earlier. A decade on from the launch, everything was on a bigger scale and much, much more complicated. The business structure needed to be simplified, but this wasn't something that could be done overnight.

The adverse comments that had followed the publication of its 1994/95 results made Vodafone reconsider its international strategy. First, it invited a dozen leading analysts to join it on a ten-day trip to see some of the international assets. Visiting the operations in Malta, Athens, Hong Kong, Fiji, Sydney and Johannesburg, the analysts were all impressed – not just by the businesses themselves, or by the surprising appearance of crocodiles on the golf course in Fiji, but also by the calibre of the local managements and the two senior executives, Julian Horn-Smith and David Channing-Williams, who'd accompanied them halfway across the world and back.

The second phase of the operation involved continuing to strengthen the company's international presence, not by moving into new markets but through the acquisition of additional equity in the existing companies. The strategy was to attempt to convert investments to 'associates' and associates to subsidiaries. That meant taking a sub-20% holding to 20% or above, and converting minority stakes into majorities.

Although Vodafone had hit its 55-million target by this time, this course of action was generally remarkably easy to achieve. Many of the original consortia included shareholders who had no interest in long-term involvement in the telecoms market and were looking for an exit, so Vodafone rarely had to pay an excessive price. Vodafone also had to be applauded for its timing – it invariably managed to convert a company to associate or subsidiary status at exactly the time that business first became profitable.

The focus on deepening the roots of existing businesses rather than adding to the number of territories left rival AirTouch with a clear run at most of the markets it fancied. But it too was feeling that there were fewer and fewer really good opportunities on offer, and during this time added only two new markets, in Poland and Romania.

The last licence that Vodafone won during this phase of its development was awarded by the Egyptian government to a consortium it led called ClickGSM. It was a pretty good opportunity – Egyptian mobile penetration was just 0.1% and there was just one competitor, the ECMS consortium consisting of France Telecom teamed with Orascom Telecom, a subsidiary of a large Egyptian industrial conglomerate. Vodafone had 30% of the shares in the consortium and AirTouch a further 30%.

Gerry Whent retired from Vodafone at the end of 1996 having, in the words of Ernie Harrison, made 'an outstanding contribution to the telecommunications industry' and developed the Vodafone Group 'into one of the world's most successful mobile telecommunications businesses'.[44]

By 1996 the established UK cellular market, with its multitier structure of network, wholesaler and retailer, had become more – overly – complicated. The many competing service providers had compounded the problem by going 'off-piste' with the tariff card, changing this or that in an attempt to get an edge. There were numerous different versions of the standard packages on offer – and, at the time, no price-comparison websites to help.

The new entrants didn't face the same problem and this was proving something of an advantage. Their licences didn't require them to use service providers, so they were in complete control of every aspect of their service. An Orange customer, for example, had a single range of packages to choose from, was given an Orange-branded phone and received a bill from Orange. Pretty straightforward.

By contrast, a Vodafone customer would have numerous different tariffs to consider, once they'd made the choice between analogue and digital; then they'd need to choose between dozens of different handsets from Motorola, Nokia, Ericsson and other vendors. Turning the handset on would bring up the word 'Vodafone' on the display but that wasn't where the bill came from. Vodac? Is that the same thing as Vodafone? But who are – is – Martin Dawes? Peoples' Phone? Talkland? Fairly complicated.

It wasn't until 1996, after Gerry Whent had retired and Chris Gent had taken over as managing director of the Vodafone Group, that this issue was properly addressed. Terry Barwick, the company's PR director, was the first to say the unthinkable: 'Can't we just call them all Vodafone?' Although not everyone liked it – though it is hard now to understand their objections – the change was made.

A further issue was that by this stage it had become apparent that Orange, with its tariffs undercutting Vodafone's, was beginning to capture some of Vodafone's higher-spending business customers. By the beginning

---

44 1997 Vodafone Annual Report.

of 1996, Orange had nearly as many customers as Vodafone had on its digital network, even though Vodafone had been running for nearly two years longer. That provided a much-needed wake-up call, and new tariffs, much closer to Orange's, were finally introduced in April 1996.

It was a very costly delay, as by then Orange was well established and remained a competitive threat for many years to come. It was one of the rare occasions that Vodafone failed to respond quickly enough to changing circumstances.

CHAPTER 9

Preparing for 1998

With both of the pacesetters – Vodafone and AirTouch – taking a break in 1996, some of the European PTTs took the opportunity to make hay in the field of third and fourth licences.

Over the next few months, France Telecom added to its licences in France and Belgium by winning the third licences in Denmark and Italy. For this latter bid, it teamed with Deutsche Telekom. BT (interestingly, not Cellnet) was the sole bidder for the fourth licence in Italy and Germany, and therefore managed to win both. All three companies also acquired spectrum in the Netherlands.

These successes merely serve to underline the remarkable lack of interest in mobile shown by Europe's leading PTTs during the industry's first decade. All had been making bullish noises about international markets, but up until 1996, none had made any real effort to establish a business in the fastest growing part of the European industry. The smaller operators weren't much better positioned. Some of them – most notably the Scandinavian trio of Telecom Finland, Norway's Telenor and Sweden's Telia – managed to rectify their failure by winning licences in Eastern Europe or farther afield, but most did not.

Why were the PTTs so unconvinced by the new service? Psychologists talk of cognitive dissonance, and that may be an explanation. The management in all of Europe's PTTs – indeed, of every incumbent phone company in the world – had grown up believing in the value of the fixed networks they were building, and they couldn't bring themselves to believe that that could possibly change. So they persuaded themselves that mobile would only ever be a marginal thing, because they couldn't cope with the idea that it might ever be more than that.

There was plenty of evidence to the contrary, however, and the market was growing daily. The introduction of competition was one of the main reasons for this, but there were other factors that were arguably more important.

The first was the arrival and maturing of digital technology. Analogue networks didn't make best use of the available radio spectrum and could only support a limited number of subscribers. The combined total of all analogue cellular subscribers had approached, but never quite reached, the 100-million mark; after 18 years, the technology had outlived its usefulness and finally been rendered obsolete.

The rapid rise in the number of connections was also helped by the declining price of handsets. The days of $2,000 units were long gone, and as the century drew to a close, entry-level phones were becoming available at prices within the reach of large parts of the world's population. Nokia – the market leader in handsets in the late 1990s – saw its average selling price drop by over 20% between 1997 and 2000 (from €218 to €169) at the same time as it was moving upmarket, adding numerous new top-end devices to its range, including the 9000 Communicator.

Finally, there was the launch of prepaid-billing arrangements (often called pay-as-you-go) in the mid-1990s. This opened up an entirely new segment of the market – that significant proportion of the world's population who weren't creditworthy but who, ironically, would most benefit from access to reliable communications. Prepaid billing made mobile service affordable for almost anyone, irrespective of circumstance. It also made it possible for mobile users to acquire a second or even multiple SIM cards – in markets where coverage was difficult, customers sometimes acquired multiple SIMs to maximise their chance of obtaining a signal.

For operators, prepaid had several material advantages over the traditional contract relationship: the risk of bad debt was zero, the cost of acquisition was close to zero, customer-care costs were low, and tariffs were substantially higher. This, obviously, led to higher margins and, crucially, much-improved cash flow, as the revenue was received up front, ahead of any costs being incurred.

The one disadvantage was that the operator didn't have a contractual relationship with the customer, so loyalty couldn't be assumed and churn (the number of people moving on) was consequently higher.

The concept of prepaid seems to have its origins in the USA: a patent for a prepayment mechanism was filed with the US Patent Office on 16 November 1994 by Andrew Wise. Ironically, uptake in the US was lacklustre, and Europe was the first part of the world to feel the impact of the new development. Telecomunicações Móveis Nacionais (TMN), the mobile arm of Portugal Telecom, was the first company to launch a prepaid tariff in Europe, and the effect on the business of the new tariffs was dramatic. The company had ended 1994 with 85,000 mobile connections; a year later, the number was 171,000; 1996 saw an increase to 332,000; and 1997 that number leapt to 762,000. During 1998 TMN's total customer base reached 1.43 million, over 80% of which used prepaid billing.

Within a couple of years, almost every other operator in Europe had followed TMN's lead. After a decade of mostly monopoly operation (from 1981 to 1991), Europe had a total of 4.4 million mobile subscribers. By the end of 1995, the total was up to 23.7 million, and by the end of 1999, 63 million people connected through prepaid tariffs and a further 112 million on contracts.

Today, although the USA still lags in prepaid numbers with fewer than 15% of all connections billed through this method, around 55% of all European connections are prepaid. In Asia the ratio is over 75%, despite the almost total absence of prepaid in Japan. Latin America is 90% prepaid and Africa is over 95%. The global average at the end of 2018 was around 75%, or about 5.9 billion people.

Deutsche Telekom may have missed the bus as far as the early mobile licences were concerned, but it managed to spot another big opportunity and took full advantage of it. Prior to the collapse of the Berlin Wall in 1989 and the subsequent collapse of the Soviet Union, even the best telephone networks in Eastern Europe suffered from decades of under-investment, and there were large numbers of towns or villages with no phone service at all. In Silesia, an area of Poland that bordered the southern part of the former East Germany, for example, there was a population of about 8 million but barely a single phone – the Soviets had thought that the strong Germanic element in the region made the Silesians troublesome, so had prevented them from communicating with each other by denying them access to telephones.

When Deutsche Telekom took on the responsibility of providing modern communications to the 16 million new German citizens in the New Federal States (those in the former East Germany that were reincorporated into the Federal Republic of Germany on reunification in October 1990), it discovered that there were more than 2,000 communities without a single phone, and that most of the switching and transmission plant dated back to the late 1930s. The company's management concluded that just to bring the network up to the most basic modern standards, an investment of at least DM20 billion (about €10 billion) would be needed. They resigned themselves to the fact that this expenditure would never generate an adequate return ...

Most of the other countries were in a similar sad position. The Hungarian telephone company Magyar Telekom (Matav) was one of the few that admitted it. Its 1988 Annual Report included a set of statistics that showed that in every respect, it fell well below the developed world standards to which it aspired, with far lower overall phone ownership, a network that experienced more line faults per month than a modern operator experienced in a century, and waiting lists of seven years in Budapest and a shocking 14 years elsewhere.

The ending of communist rule gave rise to many investment opportunities in the newly liberated countries although at first not all operators were equally enthusiastic. US West was one of the first to become active in the region, when it applied for several of the new regional cellular licences in Russia. By 1995, it had stakes in cellular businesses in Hungary, the Czech and Slovak republics, and, of course, Russia. These holdings were complemented by several key investments in Western Europe, including a 50% stake in the UK's Mercury one2one, and a 26.8% interest in TeleWest, one of the largest of the UK's new cable TV companies. Among the Bell companies, this portfolio of international assets was probably only bettered by PacTel's collection.

US West chose pretty well, selecting Telecommunications Inc (the largest US cable company) and Cable & Wireless as its respective partners in its television and mobile ventures in the UK, Bell Atlantic as its partner in the Czech and Slovak republics, and the local phone company Matav as its partner in Hungary. Matav would itself soon come under the control of a joint venture between Ameritech and Deutsche Telekom, the company that was to invest the most in the revival of the phone systems in these countries.

Apart from its considerable investment in Hungary, Ameritech's interests were confined to an equal share of a 49% stake in Telekomunikacja Polska, Poland's national phone company, in partnership with France Telecom. This, at least for the moment, was the French company's only investment in Eastern Europe.

Deutsche Telekom, by contrast, had looked to establish a strong presence in the region almost immediately. Having perhaps punched below its weight in the opening rounds of the competition for western Europe, it was out to rectify that in the east. In 1992 it became one of three foreign investors in Ukraine's first cellular operator, an NMT-450 business called Ukrainian Mobile Communications. In 1993 it became an early investor in Russian cellular, taking an initial 23.5% shareholding in Mobile TeleSystems (MTS), a company that would become one of that country's big three cellular operators. In the same year, it entered the joint venture with Ameritech in Hungary, acquiring a 30% interest in Matav. Investments in the second cellular companies in Poland, the Czech Republic and Slovakia were to follow over the next two to three years, then a 35% stake in Croatia's Hrvatski Telecom.

Later, its Hungarian subsidiary Matav was also encouraged to expand in the region. It acquired a 51% stake in an entity that controlled Macedonian Telecom in 2001, following that in 2005 with 51% of Telekom Montenegro.

The second licences in Eastern Europe gave some of the other PTTs a chance to expand their footprints. One of the first came in 1993, when Telecom Finland (later Sonera Corporation), together with the Cukurova Group, Ericsson and some local investors, established Turkey's first GSM operator, Turkcell. This astute move provided Sonera with a springboard for further investment in Moldova, Georgia, Azerbaijan and Kazakhstan.

The following year Telenor, Telia and the Dutch PTT Telecom (later Koninklijke KPN or Royal KPN) all got in on the act, becoming partners with Sonera in the second mobile licence in Hungary. Telia and Sonera worked in partnership in the Baltic region, acquiring control of mobile companies and their PTT parents in Latvia, Lithuania and Estonia between 1994 and 1998. Telenor, meanwhile, acquired companies in Montenegro, Ukraine, Serbia and Bulgaria between 1996 and 2013.

Towards the end of the 1980s, the European Commission had had an idea. In a rare moment of genuine insight, it had concluded that, ideally, all European telecom markets should be deregulated and entirely open to competition. Such a fundamental change would obviously require extensive forethought and preparation on the part of the Commission, not to mention the companies themselves, so the bureaucrats set a date well in the future for the change: 1 January 1998.

As the 1990s wore on, it gradually dawned on the governments of Europe that an organisation might be in a better position to cope with commercial competition if it were itself a commercial entity; and perhaps even privatised. Back in 1980, the idea that a telephone company should be privately owned was an unfamiliar one. To think that it might be possible to convert such a beast into a successful commercial venture was to think the unthinkable. Yet it had happened to BT in the UK and that appeared to be thriving. Around the world, management of other, similar, companies began to sit up and take notice.

The transformation of the British company had been remarkable. Immediately prior to privatisation, BT had employed an astonishing 243,000 people. This was way too many and BT knew it, but as a state-owned business, little could be done about it – any rise in unemployment was always politically sensitive and, of course, there were the unions. It was no surprise that took BT a while after privatisation to begin addressing this difficult issue, but when it did, the results were gratifying.

The company offered a small voluntary redundancy programme in 1991 and a few thousand of its workers responded. Then, in mid-1992, the company launched 'Release 92' – one of the largest redundancy programmes ever seen in the UK, with 29,300 workers being let go on a single day! Business went on as before and the quality of service was maintained at the same, adequate, level. A further 30,000 employees left over the next two years, bringing the total workforce down to about 135,000.

The lesson wasn't missed in Europe, and one by one the large PTTs began to ready themselves for privatisation, ahead of the 1998 deregulation deadline. Governments brought together teams of lawyers, bankers, brokers and PR consultants in an attempt to show off their companies in the best possible light. One of the key techniques was to control the flow of information from the companies to their prospective investors, which often involved hiring just about everybody who might possibly have an opinion, to prevent them from expressing that opinion.

The November 1996 privatisation of Deutsche Telekom is a good example of this. The prospectus listed those involved in the issue, which included three 'global co-ordinators' – Deutsche Bank, Dresdner Bank and Goldman Sachs – and no fewer than 40 'regional managers' which read like a who's who of the financial services industry and included Lehman Brothers, NatWest, Nomura, Paribas, Robert Fleming, Rothschild, Swiss Bank, Warburg et al.[45] Representatives of these organisations would attend meetings with the company and be told why everything was wonderful. Soon there would be thousands of pages of near-identical, extremely positive analysis hitting the desks of institutional investors around the world, with the same simple message – buy!

At this stressful time for the industry, there was at least one apparent consolation: everyone save BT was in the same boat. This perception held

45 Deutsche Telekom AG, offer of 113,000,000 Ordinary Shares (nominal value DM 5 per Share), 17 November 1996.

sway for a while – a few years, in fact – but by the middle of the 1990s, it was clear that not everyone was in the same boat – the people at Vodafone and AirTouch seemed to be in powerboats, while the rest of the industry was stuck in leaky dinghies.

Suddenly, 1998 seemed very close at hand, and everyone was looking to buy something, anything, that might help offset the almost certainly dire effects of competition. Throw into this mix a growing amount of hype about the possible impact of the internet, and the vast expenditure on the Y2K 'millennium bug' (a problem in the coding of computerised systems that was projected to create havoc in computers and computer networks around the world at the beginning of the year 2000) and conditions became just right for a late-20th-century gold rush.

All around the world, stock-market valuations of companies with even the slightest exposure to tech, telecoms or electronic media began rising.

CHAPTER 10

The hunting season

When Chris Gent took over as managing director of the Vodafone Group, the network had been running for 12 years. Vodafone had come a long way in a pretty short time; now it needed to see exactly where it was and where it might go next.

To help, Gent called in a leading firm of management consultants and representatives from its financial advisors, and the conclusion was unanimous and unambiguous: Vodafone had two options, hunt or be hunted. The company had to either get bigger through a series of acquisitions or prepare to be swallowed up by some larger competitor.

If the 1980s had been characterised by deregulation and increased competition, during the 1990s the pendulum was beginning to swing back. At the start of the decade, three giant transatlantic alliances had been formed to address the needs of the corporate market: AT&T had allied itself with Unisource, a consortium of four European companies (KPN of the Netherlands, Swisscom, Telia from Sweden and Spain's Telefónica); France Telecom and Deutsche Telekom had created Global One in partnership with Sprint, the third-largest US long-distance operator; and BT had teamed up with MCI to form Concert.

Then, as the decade progressed, the Bell System began to reassemble itself. In 1995 Bell Atlantic and NYNEX merged their cellular interests to create a business with a single unified network stretching all the way down the east coast from New England to Washington. The following year, the two parent companies announced that they too intended to merge.

The regulatory bodies looked at the deal – and realised there really wasn't any way they could stop it. Since the 1984 break-up of AT&T had created seven separate companies, each with its own specific territories, the seven didn't really compete with each other, so a merger couldn't be deemed to be anti-competitive. Despite vigorous protests from AT&T and MCI, the merger was eventually allowed to proceed. The $25.6-billion deal created a company with over 40 million phone lines, well over a quarter of the national total.

By then, a second deal of this kind had been announced and completed. PacTel's decision to spin off AirTouch had been appreciated by the financial markets, but the west coast business was now rather smaller than some of its peers. SBC saw the opportunity and proposed a merger. Once again, the argument prevailed that the two operated in separate regions with no overlap. SBC completed its $16.5-billion deal in April 1997. The acquisition was the third largest in US corporate history and created the second largest telecoms company in the US, behind AT&T.

In May 1998 SBC announced that it intended to merge with Ameritech, the Bell company that served Chicago and the Midwest. The proposed deal would create an absolute monster, the country's largest phone company, with some 57 million phone lines and a market value of $62 billion. Would the regulators allow it? Could they stop it? They couldn't – and in October 1999 the deal was completed.

By this time Vodafone's valuation had risen to around £15 billion, nine times the £1.7-billion flotation price, but it was still vulnerable to the threat of a takeover. While some of the American operators, or indeed the larger European PTTs, had established a reasonable presence in mobile, none of their fiefdoms came close to matching the empires created by Vodafone or AirTouch, which between them had something close to continental coverage in Europe. Mobile was by far the fastest-growing part of the telecoms market and none of the big boys felt they had a sufficient exposure to it.

The Europeans were probably less of a threat – most were too preoccupied with the deregulation of their market to have enough time to consider acquisitions. But the four Bell companies, AT&T, and the new rising star, WorldCom, were all potential predators.

WorldCom had begun life in 1983 as Long Distance Discount Service (LDDS); in 1995 it changed its name to WorldCom; and by 1997 it had become the fourth-largest long-distance company in the country, behind AT&T, MCI and Sprint. It had no intention of being fourth for long, and in October that year it made an all-paper bid for MCI. BT, which owned 20% of MCI, counter-bid, as did GTE, but the CEO of WorldCom, Bernie Ebbers, wasn't to be denied. He eventually paid over $30 billion for MCI, all of it in the form of WorldCom stock, as well as assuming a further $5 billion in debt.

AirTouch was both a potential predator, but also a target. It had majority stakes in mobile businesses in Portugal and Sweden and significant investments in networks in Belgium, Germany, Italy, Spain, Poland and Romania. Vodafone had the UK, France, Germany, Greece, the Netherlands and Sweden. The only place the two competed was in Germany, but there would be plenty of ready buyers for that. It was an almost perfect fit.

Gent thought the best option would be to suggest a merger between it and AirTouch's European assets. In early 1998 Gent, Julian Horn-Smith and some key advisors flew to America to test the water. Working from east to west, the Vodafone team first met Ray Smith, the boss of Bell Atlantic; he was cordial, but it became apparent that Smith also liked the look of AirTouch – and not just its domestic assets. Unhelpful.

WorldCom's Ebbers had been included on the itinerary as a back-stop; if Vodafone had to launch a hostile bid for AirTouch, it would need to line up an agreed buyer for the domestic assets. But that meeting wasn't much more productive – Ebbers told Gent that he was going to move into the mobile market 'next year' and that he too had his eyes on AirTouch ... all of it. 'The company's name is WorldCom', he reminded them.

Finally, the team arrived in San Francisco to meet AirTouch's chief executive Sam Ginn and the company's chief operating officer Arun Sarin. No agreement was reached at that meeting nor at another later in the year. Ginn listened carefully to the proposal but didn't like it. The two agreed to disagree and went their separate ways.

Bell Atlantic's decision to bid for AirTouch on 1 January 1999 brought Vodafone's options into sharp focus. On 3 January Vodafone published a counter-offer: a swap of Vodafone shares for AirTouch's, plus some cash, giving a total value of $56 billion, nearly 25% above Bell's number.

Before Bell could respond, Bernie Ebbers at WorldCom entered the ring. He suggested a $55-billion stock swap to AirTouch. Although this didn't top Vodafone's bid, Ebbers expected his company's shares to rise on the back of the news, thereby taking the value of the WorldCom bid above Vodafone's.

Exactly the opposite happened. When Bell first bid, its shares were marked down by a couple of per cent. When WorldCom bid, its stock dropped by 12% in just two days, making the bid worth less than $49 billion. Vodafone's bid had been greeted with rising prices all round, so it was clear what the markets thought the best option was. AirTouch's shares were obviously going up, but the other moves spoke volumes.

A subsequent rise in Bell Atlantic's shares suggested two things. First, the market didn't believe it would beat Vodafone; and, second, that Bell Atlantic and AirTouch weren't a combination Wall Street particularly liked.

By contrast, Vodafone had been given a ringing endorsement from the financial community which – crucially – increased the value of the bid for AirTouch shareholders. The close relation with shareholders, established over many years, had paid off spectacularly.

But Ginn was ambivalent: it was the deal he'd wanted, but the wrong people were in the driving seat; it should have been the other way around, he thought, with AirTouch buying Vodafone. Nonetheless, when Gent threw another $3 a share into Vodafone's offer, the deal was struck.

Ultimately, it didn't just come down to just money. First, Vodafone was focused, entrepreneurial and dynamic – everything Bell Atlantic wasn't. Second, Bell Atlantic was essentially a traditional telephone business, with over 40 million fixed-access lines; mobile would have been very much the junior partner. Third, the 1993 spin-off from PacTel had created a sense of liberation within AirTouch and no one welcomed the prospect of becoming part of the Bell System again. And finally, Vodafone came with Chris Gent; as Arun Sarin later acknowledged, 'We'd never have sold our business to a company that didn't have Chris Gent in it.'[46]

The new Vodafone AirTouch was a giant. It had a presence in 23 countries, 29 million customers, annual revenues of £6 billion and cash flows of over £2 billion. It was the world's sixth-largest telecommunications company and the largest mobile operator, with 8.5% of the globe's mobile subscribers. Its stock market valuation exceeded $110 billion, making it the third-largest company listed on the London Stock Exchange.

But now Gent had to tread carefully. The team at Bell Atlantic weren't happy with the outcome of the bid. AirTouch and Bell Atlantic were partners in Prime PCS, whose bidding strategy at the 1995 auctions meant that it filled in most of the major gaps in the two companies' coverage, and as a result, the two companies' mobile businesses were entirely complementary. There was a strong case for bringing them together under common control, to counter the growing power of AT&T's mobile operations.

The situation was made more complicated by Bell Atlantic's recent decision to merge with GTE. While not quite in the same league as AirTouch, GTE had nearly 5 million mobile customers (compared with 8 million at AirTouch and 6 million at Bell) and owned licences in several of the large markets Bell Atlantic lacked, including Los Angeles, San Francisco, Dallas, Houston and Cleveland. At a push, Bell Atlantic could just walk away from the table if the right deal wasn't forthcoming.

The sticking point was, inevitably, the issue of who would control any merged mobile business. Sam Ginn, who was leading the talks for Vodafone, thought it should be AirTouch; the chairman of Bell Atlantic, Ivan Seidenberg, felt it should be his company. The talks stalled and it came as no great surprise when Bell Atlantic announced that it intended to dissolve the Prime PCS partnership.

That left Vodafone with a problem. Although it was a powerful force on the west coast, it had no presence in a number of key markets, including New York, Chicago, Philadelphia and Washington. There was no chance of creating a national service without a licence in New York and Chicago, and most corporates would probably require the other two markets as well.

---

46 Personal communication with Arun Sarin.

There appeared to be three options. One was to buy some of the smaller independent PCS operators, such as Omnipoint in New York – though filling in all the gaps would be both tedious and expensive. The second was to strike a deal with the FCC. The collapse of some of the C-block licensees, such as NextWave, had left a fair amount of spectrum in limbo, but again, piecing together a national footprint from these fragments would require extended negotiations with a party that had the clear upper hand. Third, Vodafone could attempt to placate Bell Atlantic and revive the partnership.

Almost immediately, the option of piecing together a national footprint by buying some of the smaller independent operators disappeared, when a small west coast operator called VoiceStream PCS bought New York's Omnipoint and Chicago's Aerial, creating a near-national GSM footprint.

The second option – buying spectrum off the FCC – was rejected as too complex and likely to take far too long to consummate. By the time Vodafone had acquired all of the necessary franchises, it would be coming in as a late fourth, fifth or sixth market entrant. That was not an attractive prospect.

So Gent went back to Bell Atlantic and laid his cards on the table. 'This isn't a merger of equals, we understand that. We'd love to have a branded entity here, but we can't see a way to do that nationally. Failing that, we'd like to be an active minority shareholder [in the enlarged group] ... but you've got to give me a premium to justify the deal to my shareholders.'[47]

Seidenberg was happy to do that, and in September 1999 the new company, initially called Bell GTE but subsequently Verizon Wireless, was formed. Vodafone contributed about 39% of the assets and emerged with a 45% stake in the business, giving Chris Gent the premium he'd asked for. Seidenberg got the national footprint he needed to compete with AT&T Wireless: the new company was more than twice as large as AT&T, with a presence in 49 of the top 50 markets and 96 of the top 100; it covered over 250 million of the country's 280-million population, including a significant proportion of those living in rural areas.

---

47 Personal communication with Chris Gent.

The partnership with Bell Atlantic wasn't Vodafone's only collaborative venture. It owned 34.8% of Mannesmann Mobilfunk, with Mannesmann holding the balance, and 21.8% of Omnitel Pronto Italia, where Mannesmann had just raised its stake to over 55%.

Chris Gent had met Klaus Esser, Mannesmann's new CEO, back in 1998 before the AirTouch takeover. The two ran the two largest independent mobile operators in Europe and Gent had thought it was important that their businesses' strategies were closely aligned. Following the AirTouch deal, it wasn't so much important as essential.

A lawyer by training and an intellectual by inclination, Esser had been appointed to the top job at Mannesmann in 1998. Hugely ambitious, he wanted to complete the company's transformation from a dull rust-belt engineering company into an international communications powerhouse. By 1998, the process was well under way, though by no means complete.

Mannesmann had become involved in telecommunications when PacTel had selected it as its partner for the German D2 digital cellular system licence bid in 1990. And when preparing its bid for the Italian licence, AirTouch had offered Mannesmann a share of the equity in Omnitel. Both Mobilfunk and Omnitel had prospered – the German company had been profitable after just three years and cash flow positive after five, while the more talkative Italians had allowed Omnitel to reach the same landmarks in two and three years respectively, which was enough to convince even the most conservative of Mannesmann's directors that the telecom business had a future.

From the mid-1990s onwards, thanks in large part to this performance and Klaus Esser's powerful advocacy, the company directed most of its substantial resources into telecommunications.

In January 1999 Chris Gent and Julian Horn-Smith met Esser and his chief strategist, Kurt Kinzius. Esser told Gent that he saw Vodafone as the main threat to his independence; Gent replied that while he believed the two assets 'belonged together', he had no interest in forcing the pace and was happy to maintain the status quo. Esser didn't believe Gent's assurances and left the meeting determined to secure the independence of his company. He intended to make Mannesmann so big that it was unassailable, even if that meant jeopardising the relationship with Vodafone.

The press and financial community didn't help. At an analysts' meeting six months later, Gent was asked, 'Now you've bought AirTouch, will you be going after Mannesmann?' Gent's reply was quite clear: 'They are our partners. We are happy with the arrangement, but we have told them that if they come into the UK market, then they had better watch out.'[48]

The Vodafone AirTouch deal had had one rather curious side-effect. The giants of the European telecoms industry looked at the new industry leader and all began to suspect that they weren't quite large enough for the new world order it represented. Vodafone's policy of acquisitions seemed to point the way, so almost everyone was on the lookout for suitable targets. But there weren't many easy options, as very few companies were selling.

Demand for mobile assets far outweighed supply, and throughout 1999 the asking price increased at a startling pace. In July 1999, for instance, BT paid $1,175 per pop, or over $2,600 per customer, for the 40% of Cellnet it didn't already own – nearly 25 times the valuation of the original Vodafone IPO just a decade earlier. The following month, Deutsche Telekom bought one2one from the Cable & Wireless/MediaOne joint venture for $4,485 per customer. In October, France Telecom bought Vodafone's stake in E-Plus in Germany for $4,900, and later that month offered to buy Veba's controlling stake in the same company for more than $5,700 per customer.

In this feverish climate, almost anything Mannesmann chose to buy would appear expensive.

In late October, Gent heard that Esser was attempting to buy a controlling stake in Orange from Hutchison. Gent immediately called Esser and said, 'Klaus, I hope you're not going to do anything without discussing matters with us first; we are your partners.'

'I'm not in a position to discuss this at this time but I'll let you know if we're going to do anything,' Esser replied.[49]

The next day he signed a deal with Hutchison to acquire Orange.

Literally overnight, a central part of Vodafone's strategy had been

---

48 Chris Gent at the Savoy Hotel, London, July 1999.
49 Personal communication with Chris Gent.

demolished. It could no longer guarantee that its key German and Italian properties would be managed in a way that was in its (Vodafone's) best interests. Indeed, it was possible that the opposite might be the case: Mobilfunk, Omnitel and Orange might act together in a way that threatened the profitability of Vodafone in the UK and elsewhere in Europe. Anything, it seemed, was now possible.

And it wasn't just Vodafone that was affected. Almost all of the continent's large telcos had extensive, uncontrolled interests that were now in jeopardy. Klaus Esser had changed the face of the industry forever.

Mannesmann had paid a huge price for the business. Including Orange's debt, Esser's offer had valued Orange at around €21 billion, 40% of which was in hard cash. This equated to about $9,200 per customer, or roughly 30 years' revenues, an unprecedented price.

As the details of the deal emerged, it became clear that Esser had laid the groundwork for his move with great attention to detail. He'd secured an irrevocable undertaking from Hutchison that the offer would be accepted, even were a higher price to be offered by a counter-bidder. As the Hong Kong conglomerate owned 45% of Orange's shares, a realistic counter-bid was almost impossible. Esser had also hired almost all the leading German banks and brokerages, PR agencies and law firms in an attempt to ensure that there was no one Vodafone could use to present its case to Mannesmann's German investors. And the timing of the move was also immaculate – only days earlier, Vodafone had sold its stake in E-Plus to France Telecom, leaving it with no route back into the German market.

Chris Gent thought he'd have one more attempt to persuade Esser to negotiate. (Up until the point a deal is declared 'unconditional' it can still be contested or abandoned.) In November 1999 Gent and Julian Horn-Smith flew to Mannesmann's headquarters in Dusseldorf and proposed a straight merger, but Esser rejected the offer. He thought Vodafone was wrong to concentrate on mobile rather than an integrated fixed/mobile offering, and suggested that the British company didn't understand the market for mobile data and had overlooked the potential of the internet. He concluded with the assertion that, 'Your shareholders can't afford to pay what my company is worth.'

'Look, Klaus, it's not your company, it's your shareholders' company,' Gent countered, 'and if you don't wish to discuss some kind of a merger, we're going to have to take the matter to your shareholders.'[50]

Shortly afterwards, Vodafone launched an unsolicited bid for Mannesmann. It was offering 53.7 of its own shares for each Mannesmann share, for a total of €124 billion. This was a huge 76% premium over the €143.5 closing price of 21 October 1999, the day after Mannesmann's offer for Orange had been announced.

In the press release that accompanied Vodafone's offer document in November 1999, Gent explained, 'We made a friendly approach to Mannesmann last Sunday. That proposal was rejected and we have now received a letter from Dr Esser making it clear he has no interest in a constructive negotiation. As a result we have decided to make an offer directly to Mannesmann's shareholders. This is the only way we can present Mannesmann shareholders with the option of investing in the world's leading, international mobile telecommunications company. I am convinced that a combination of Mannesmann and Vodafone AirTouch will produce enhanced growth prospects and superior value for the shareholders of both companies. I hope that the shareholders of Mannesmann will accept our all share offer which we believe to be in the best interests of the shareholders of both Mannesmann and Vodafone AirTouch.'[51]

According to Esser, however, Vodafone's offer was nowhere near enough and still undervalued the business substantially. He advised his shareholders to refuse it.

The following week, on 22 November, the offer for Orange became unconditional, and Hutchison Whampoa, Vodafone's old rival, became the German company's largest shareholder. Esser took the opportunity to announce that he was accelerating plans to demerge the telecommunications businesses from the group, something investors had been pressing him and his predecessors to do for several years.

For the next couple of weeks, the two companies both tried to emphasise

---

50 Ibid.
51 Vodafone press release, 19 November 1999.

the merits of their respective positions. Chris Gent took the unusual step of writing to the Mannesmann employees, assuring them that Vodafone's intention would be to grow the business – this wouldn't be an old-fashioned asset-stripping raid. Esser brought his team to London, where he held a briefing for analysts and journalists at which he preached the gospel of fixed/mobile integration, but the message was somewhat diluted by the presence of Hans Snook, Orange's chief executive, who was very firmly in the 'mobile-only' camp.

Vodafone responded to Esser's comments by noting that it too believed that mobile voice, data and the internet were all areas with substantial potential, but, it asked, 'What can Mannesmann do on its own, that it could not do better as a part of Vodafone?'[52] The lack of any kind of answer to this was to influence a number of investors over the coming weeks.

And Vodafone had a key advantage: over the years, Chris Gent and Julian Horn-Smith had taken the time to meet and get to know the financial community. These investors knew what to expect from Gent and Horn-Smith and knew that they would deliver on their promises. Esser, by contrast, had remained aloof, so for many fund managers, he was something of an unknown quantity – and financial markets hate unknowns and uncertainty.

Vodafone published a formal offer document on 23 December 1999, entitled *Better Together*. It highlighted the interlocking nature of the two companies' mobile assets, but also the differences between the two, all of which suggested the British company was the better bet: Mannesmann only operated in a few European countries, while Vodafone had a global presence; Mannesmann was a conglomerate, Vodafone was focused on telecommunications. With an eye to the workforce and politics, Vodafone promised that there would be no redundancies in Germany and re-emphasised its desire to invest in the country. And, once again, it asked that difficult question – 'What can Mannesmann do alone that it cannot do better together with us?'[53]

As the new century began, the financial markets started to move in a

---

52 Vodafone press release, December 1999.
53 *Better Together*, published by Vodafone on 23 December 1999.

way that suggested the British company was now getting the upper hand in the struggle. Vodafone's share price began to rise, steadily closing the gap between the value of its offer and the Mannesmann share price.

At the end of January 2000 the two sides met in Paris, where Gent and Horn-Smith told Esser that Vodafone was about to announce a new joint venture with Vivendi. It was a deal that Esser had been looking to do with Vivendi himself, and he now had to accept that this lifeline had gone. Although his position was hopeless, Esser fought on – he reiterated his view that Vodafone was offering too little, so Gent agreed to a small increase in the price, and various other concessions, including employment levels and investment prospects in Dusseldorf, were negotiated.

On 10 February 2000 Vodafone announced the deal. It was a truly historic moment. Not only was this the first time a major German company had fallen to a hostile bid, at a valuation of $160 bn, it was also the largest-ever takeover in corporate history.

In less than 20 years, from a standing start and without any of the advantages that the giants of the industry enjoyed, Vodafone, a small start-up from the Thames Valley, had changed the face of telecommunications and indeed the entire business world. It had taken on and beaten all comers, and was the clear and undisputed leader in the global mobile market.[54]

---

54 Chris Gent suffered a fair amount of abuse in the weeks after the takeover, from both the popular press and shareholders. But it's one thing to be called names, quite another to face trial, which is the fate that befell poor Klaus Esser. Accused of having allowed Mannesmann to be acquired merely to line their own pockets, Esser and five others were charged with 'aggravated breach of shareholder trust'. In the first trial, which began in January 2004, all the defendants were acquitted. However, a higher court overturned the ruling, so the six had to face a retrial two years later. On this occasion, the court agreed to drop the charges in exchange for payments totalling €5.8 million – a fraction of the cost of the legal actions.

CHAPTER 11

Information-superhighway robbery

At the start of the new millennium the European mobile industry had a new challenge to face. The era of the mobile internet, driven by the development of revolutionary third-generation (3G) mobile technology, was about to begin.[55] This, it was claimed, would allow operators to provide an almost limitless array of new services over a network that for the first time would be genuinely universal. Varying standards would be a thing of the past and a single phone would work anywhere on the planet.

Governments were quick to realise that this profoundly positive mood could be turned into hard cash. The new UK chancellor of the exchequer, a Scot called Gordon Brown, hired an academic, an expert in game theory (the analysis of strategies for dealing with competitive situations), and told him to design an auction process that would maximise receipts, irrespective of the cost or consequences.

The auction of the universal mobile telephone service (UMTS) licences was thrown open to all comers – no question of foreign-ownership restrictions here. And the available spectrum – 160MHz of paired frequencies, plus a further 25MHz of unpaired, in the 2.3GHz band – was split it into five parcels, two of which were larger than the other three. The largest and most attractive – the A block – was reserved for a new entrant, thereby

---

55 The first-generation mobile phones were analogue (AMPS, NMT, TACS, etc.); the second generation were digital (GSM, CDMA, TDMA, etc.) and the third generation were higher-capacity digital.

ensuring that only one of the two market leaders – Vodafone and Cellnet – could get as much spectrum as it needed.

This was the world's first 3G auction, so the level of interest was enormous. The bidders included all four UK operators (Vodafone and Cellnet, one2one and Orange) and nine other hopefuls, including WorldCom, Telefónica and NTL, the UK cable-TV company. There were several dark horses, including an obscure Canadian company called Telesystems International Wireless (TIW).

The process started at 1.30 pm on Monday 6 March 2000. Round one closed with bids totalling £545 million on the table. By the end of the week, that had doubled. Ten days later, on 22 March, the first £1-billion bid was made, at which point the total stood at over £3 billion. Vodafone placed the first £2-billion bid a week later.

After 87 rounds of bidding, with £9 billion committed, not one of the participants had dropped out. By round 108, when the first £3-billion bid was made, there were eight left in the race.

It took a further 42 rounds before the rotten process was finally over. The government had netted more than £22 billion, or 'enough money to build 400 new hospitals' – had that been their intention, but it was not.[56] As it was, they squandered it.[57]

The A block was bought for £4,385 million by a consortium led by TIW, which turned out to be effectively a flag of convenience for Hutchison Whampoa, which owned all but 9.1% of the consortium. Six months after selling Orange to Mannesmann, Hutch was back in the game.

Vodafone got the large B-block licence that it wanted, but at an astronomical price – £5,964 million, or over £100 per pop.

The three smaller licences went to Orange for £4,095 million, Cellnet for £4,030 million, and one2one for £4,004 million.

The government greeted the result as a triumph. Not so the industry, however: once the euphoria of the contest had faded away, operators had

---

56 'The biggest auction ever: The sale of the British 3G telecom licences' by Ken Binmore and Paul Klemperer, in *The Economic Journal*, Blackwell Publishers, March 2002.

57 The hospitals weren't built, but the payments extorted from the industry added over £400 – or £20 per annum for 20 years – to the cost of communicating for everyone in the country.

a little time to reflect on what had just happened. It was appalled at the price it had had to pay – four, five perhaps even ten times their pre-auction estimates. And contrary to what the government and its expert claimed at the time, the operators didn't see this as a 'sunk cost' that they would never recover: they saw it as a form of legalised extortion and they recognised they had an obligation to their shareholders to get it back as quickly as they could.

And the UK was only the beginning: just around the corner, similar auctions were being organised by the governments in the Netherlands, Germany, Italy, Austria, Switzerland and Belgium. If those raised similar amounts, the largest operators would be looking at a bill of over £20 billion for 3G licences in Western Europe alone. And that was before they even started to build the new networks …

Fortunately, after the UK auction, a fair bit of the hype surrounding 3G evaporated and none of the other markets managed to reach the same price per head of population. Germany netted more (just over €50 billion), but then it had a population 30% higher than the UK's. In that country, six licences were awarded – to the four incumbents, Vodafone, Deutsche Telekom, E-Plus Hutchison (KPN's subsidiary in partnership with Hutchison) and the newcomer InterKom (including BT, with partners VIAG and Telenor); and Mobilcom (in partnership with France Telecom) and Group 3G (an alliance formed by Telefónica and Sonera).

The fall-out was immediate. Within hours Hutchison announced that it was leaving the partnership with KPN, ending the Dutch company's dreams of a pan-European alliance and bringing the nightmare task of financing both the licence and network-construction costs into sharp focus.

BT found itself with a similar issue. VIAG and Telenor exercised put options that required BT to buy their stakes in the business – at a price of more than €14 billion. BT had hoped to take control of InterKom at some stage, but not quite now! This was the end of BT as a major force in the auctions and, arguably, a major force in the mobile market.

And Mobilcom found itself in the same position as KPN. It had expected France Telecom to finance the building of the network but shortly before this auction, France Telecom had spent over £25 billion buying Orange from Vodafone and it just didn't have the cash. The alliance was eventually terminated in June 2002 in an atmosphere of mutual hostility and acrimony, and the proposed network was never built.

Rumours soon began circulating that Group 3G hadn't intended to win a licence at all! It seems it had only been trying to reduce its competitors' financial firepower ahead of some imminent spectrum auction in Brazil but had been caught short by the abrupt end to the bidding. That Group 3G didn't have a single employee in Germany and hadn't arranged finance for the project lent credence to the story, but whether it was an accident or not, it was a disaster. Less than two years later, Telefónica threw in the towel, returned the licence to the government and wrote off more than €11 billion.

In Italy the government raised only €12.6 billion for its five licences, one third the amount paid for the UK licences. It was so upset by this outcome that it refused to publish the result for five days.

The remaining auctions, in Austria, Switzerland, Norway, Portugal, Sweden and Belgium, failed to attract many outside bidders, so generally the licences fetched little more than their reserve prices.

Finally, in France – one of the most attractive mobile markets on the continent, with only three competitors and the lowest level of penetration anywhere in Western Europe – the only interest came from the three incumbents, who made it clear to the government that they wouldn't pay the €4.95-billion price tag for a 3G licence, and after a year or more of prevarication, the government agreed to cut the fee by 80%.

Within months of these auctions ending, the chief executives of British Telecom, Deutsche Telekom, France Telecom, KPN and Sonera all lost their jobs. All five companies had to undergo extensive refinancing, inevitably on unfavourable terms. All had to rein back their level of investment, and all had to dispose of assets that until then had been considered core parts of the business.

The governments of Western Europe had squeezed nearly €110 billion out of the mobile industry in just a few months. The damage this did to the industry is incalculable and guaranteed higher phone prices for all for a very long time to come.

The question has to be asked – was it worth it? The answer is almost certainly no. At the time the licences were awarded and for several years after that, there really weren't many good 3G devices available, so the service couldn't easily be accessed. The mobile operators weren't especially imaginative when it came to the creation of content. Some borrowed NTT DoCoMo's 'i-mode' mobile internet service, or tried to develop their own multimedia services, but none of these was especially compelling. 'Vodafone Live!' was probably the best of the lot, but it wasn't that special.

Three years after the auctions had ended, less than 2% of European mobile users had a 3G handset; after five years, the number was still less than 8%.

The financial markets didn't like what they'd seen and were quick to react. Although it would be overstating the case to say that the 3G auctions were the reason for the burst of the dot.com bubble – a period of excessive speculation from around 1994 to 2000, coinciding with a period of extreme growth in the use and adoption of the internet – it's a fact that stock markets around the world started to lose their enthusiasm for tech and telecom stocks on 27 April 2000, the final day of the UK's 3G auction. The markets thought the operators had paid too much for these licences and started to cut the share prices of the participants.

The situation in Europe was exacerbated by events in the US, where valuations placed on telecoms stocks on Wall Street had been steadily rising throughout the mid-1990s, and spectacularly towards the end of the millennium. Not wishing to miss the chance of a lifetime, companies began making ever-larger investments in telecoms infrastructure, laying down millions of miles of fibre in the hope of capturing a share of the new, exploding telecoms market. Investment bankers, fearful of missing the investment bonanza, became less fastidious when assessing opportunities: in the words of Citigroup director Robert Rubin, there was a new 'tendency of lenders to forego discipline in credit extensions'.[58]

Eventually, however, it became clear that the party was over. Today, almost 20 years on, these businesses and their valuations still haven't fully

---

58 Speech by Robert Rubin of Citigroup to the Year 2000 Salomon Smith-Barney
   Telecommunications Investors' Conference.

recovered.

The collapse in investor confidence dealt a further body-blow to the industry. All of a sudden, the IPO market was closed for business – no one in the investment business wanted any more telecom assets, thank you very much. The industry could no longer use its paper to finance major deals, so consolidation, which had done so much to create value during the late 90s, was no longer possible. The pain was most focused on the smaller, independent European companies who had come into the market on a tide of continental deregulation. Some were so desperate to unload assets that they were prepared to take rock-bottom prices, while others failed to manage that and were driven into bankruptcy. This, of course, had a further adverse impact on sentiment – and valuations.

In the aftermath of the auctions, the magnitude of the European industry's debt problem became increasingly apparent to outside observers. One study published towards the end of 2002 suggested that taken together, the European telcos were looking to sell assets worth upwards of €90 billion – or rather, needed to raise €90 billion from asset sales. That figure took account of numerous asset sales that had already taken place, such as BT's disposals in Japan and Spain, Deutsche Telekom's sale of its cable TV assets and its Sprint stake, a near-wholesale clear-out by KPN, and so on. The situation looked daunting.

In practice, it wasn't quite as frantic as that – at least, it wasn't for the PTTs. They knew they would still have the support of their national governments, come what may, and weren't unduly worried if the share price took a bit of a hit. Most of these organisations were still bureaucracies at heart, unused to the ways of the financial markets and perhaps, uncaring about the goings-on in those casinos. And it should be remembered that this was happening just a few months after most of these companies had been exposed to real competition for the first time. What did it matter if a few of the new competitors had gone out of business? It made life a little easier! As the small print says, 'share prices can go down, as well as up'.

CHAPTER 12

The new superpower

Klaus Esser had taught Chris Gent and his team a valuable lesson – don't expect partners to behave themselves. Take control – everywhere.

Three of Vodafone's main operating companies were already consolidated subsidiaries (majority owned, therefore controlled) – those in Germany, Italy and the UK – but the company had interests in many other markets that needed some resolution. Foremost among these were four markets that Vodafone considered strategic: France, Spain, Japan and the USA. Could they obtain control of any or all of these?

Spain had been close to the top of Julian Horn-Smith's wish list for some while. Vodafone was one of the largest shareholders in Airtel with 21.7% of the equity, but BT had another 16% and wanted to integrate the mobile company with its existing Spanish fixed-line business, BT España. The remaining shares were held by a number of local banks and financial institutions.

Vodafone talked to BT and asked it not to stand in the way. It told BT that if it managed to acquire more than 55% of the shares, BT could buy those if it wished. However, it had become clear that BT was in a bit of a mess, with too many early-stage businesses and an excessive amount of debt thanks to the 3G auctions. Under the circumstances, it was as good a deal as could be expected, and BT agreed to the arrangement. In July 2000 Vodafone bought out the remaining Spanish shareholders, bringing its stake in Airtel to 73.8%.

It now had controlled subsidiaries in four of Europe's five largest markets, an unmatched footprint.

After these deals had been completed in July 2000, Vodafone wondered whether BT would take up the offer to acquire some shares in Airtel. It did not – it was still struggling to get its finances back under control. This involved spinning off Cellnet and disposing of several other assets, including eventually its residual stake in Airtel, which Vodafone went on to acquire in June 2001.

In the meantime, Vodafone's growth was continuing at an unprecedented rate. In the UK, the company's total customer base increased from 7.94 million at the end of 1999 to 11.66 million a year later – the 3.7 million new customers equated to more than 6% of the UK population. In Italy, it added almost 5 million people to its base, or 8% of the population, and in Germany, nearly 10 million or more than 12%.

During this period, the disposal of Mannesmann's engineering, automotive and luxury clock and watch businesses raised €9 billion, and the sale of the Italian fixed-line operator, Infostrada, brought in a further €9 billion.

In late May 2000, Vodafone announced the sale of Orange to France Telecom for a remarkable £25.1 billion, £14 billion of which was to be paid in cash. The deal left Vodafone in by far the strongest position of any European operator – cash rich, with an enviable flexibility – while France Telecom had joined the long list of over-stretched former monopolies. Warren Finegold of UBS, the bankers who handled the sale, summed it up when he noted that 'this deal reversed the polarity of the European telecom industry in Vodafone's favour'.[59]

Later in the year, Vodafone turned its attention to Japan, where it had acquired stakes in ten regional operators through the AirTouch merger.

Japan is divided into ten administrative regions, three of which – Tokyo, Kansai and Tokai – are urban and account for the majority of the population, while the other seven are more rural, with much lower population densities. When mobile was first introduced to Japan in 1979, NTT DoCoMo was given a licence to operate in each of the ten regions.

---

59 Personal communication in 2003.

In 1987 and 1994 additional competitors were licensed: IDO (Nippon Idou Tsushin) in Tokyo and Tokai, and Daini Denden (DDI) everywhere else. (Later, IDO and DDI would merge with KDD, an international fixed-line operator, to form KDDI.)

In the 1990s the Japanese government issued two licences for the new PDC digital technology in each of the big three urban regions to Digital Phone and Tu-Ka; and one additional licence in each of the seven rural markets to six separate subsidiaries of a joint venture between the two new entrants, called Digital Tu-Ka.[60] AirTouch was a founder shareholder in the three Digital Phone businesses and all six of the others. (In typically Japanese fashion, none of the 12 new operating companies had exactly the same shareholding structure as any of the others. Numerous large corporations had shareholdings, including Hitachi, Marubeni, Matsushita, Nissan and Sony, along with various regional railway and electricity companies.)

Following the AirTouch deal, Vodafone had been gradually increasing its investment in the business, which by then had rebranded itself as J-Phone. In July 1999 it bought a 12% stake in J-Phone Tokyo. The next step wasn't planned: in August the Nissan car company, which was a shareholder in the nine regional companies, sold its entire holding to Vodafone for $139 million – Nissan's core business was under pressure and the company's management didn't like the look of the investment needed to establish 3G in Japan, so it decided to quit the business.

Soon afterwards, Vodafone bought Cable & Wireless's interests in J-Phone Tokyo and the other two urban operators, J-Phone Kansai and J-Phone Tokai. And in September it bought out Cable & Wireless completely, acquiring its 4.5% stake in each of the six regional businesses. Together, these two transactions cost it $550 million.

Vodafone was now much the largest foreign shareholder in the business, with 20–28% of the equity in each of the companies. However, this was still a long way away from control.

In mid-2000 the Japanese government surprised the industry when it announced that it was going to award only three 3G licences, rather than

---

60 The regions of Kyushu and Okinawa were treated as a single entity for the purposes of these licences.

four. One of the existing operators would therefore be excluded. DoCoMo and KDDI were probably safe, leaving the choice between J-Phone and Tu-Ka. In response to this, J-Phone announced in March that it was reorganising itself into three regional holding companies and taking control of the six regional businesses. A new entity, J-Phone Communications, was created to act as a holding company for these, 54% owned by Japan Telecom, with the balance being split between Vodafone with 26% and BT the remaining 20%.[61]

After several more byzantine and subtle deals, Vodafone emerged victorious as the controlling shareholder in both Japan Telecom and J-Phone. On September 2001 Vodafone acquired the majority stake it required, setting several new records along the way, including the largest-ever foreign investment in Japan.

Following the takeover, Vodafone set about the reorganisation of the business. Where once there were nine, now there was just one – one billing system, one national pricing plan, one procurement process and one set of overheads. The mobile network was re-engineered and the fixed-line activities were set up as a separate company, which was sold in August 2003.

In just three years, Vodafone had completed four of the largest corporate deals of all time. It had changed profoundly. Its 1998 Annual Report reported a base of 5.8 million customers and about 10,000 employees, most of both in the UK. The 2001 version showed a customer base of some 83 million and 53,000 employees, most in the US, Germany or somewhere other than the UK. The board had changed too, with the addition of six Americans and five Germans.

For some companies, this might have been enough. However, Vodafone remained entrepreneurial and expansive, with another four deals being completed in 2002 alone.

The first was the acquisition of a fragment of China Mobile's equity – a

---

61 Although this was a welcome simplification, it still wasn't that simple. Ownership of J-Phone Communications was itself divided between Vodafone and J-Phone Holdings Co., the latter being jointly owned by Japan Telecom and BT.

2.14% stake for which it paid $3.25 billion. This curious moved divided opinion but the stake was sold some years later for $6.6 billion, so it can't be considered a mistake. The following month Vodafone acquired 25% of Swisscom Mobile, a subsidiary of the Swiss PTT, for £1.8 billion; five years later it was sold for a similar amount to pay down debt.

At the end of 2002 Vodafone announced that it was to acquire Eircell, the mobile arm of Ireland's national PTT, at an attractive price that was equivalent to just €3,500 per customer, a third of the price Mannesmann (and France Telecom) had paid for Orange. And the last of the four deals was the $973-million investment in Iusacell, Mexico's second-largest operator, for a 34.5% interest; this didn't work out well, though, and less than three years later Vodafone sold out for next to nothing.

Vodafone had become so big so quickly that it was no longer possible to run it effectively without the introduction of more formal processes and structures. After the AirTouch deal in 1998, Vodafone had called in Bain, the American management consultancy, to review the two halves of the business and develop a modus operandi for the future. Bain had concluded that although there were superficially similarities, Vodafone was quite unlike AirTouch: there was far more flair within the UK business and a greater propensity to gamble and take risks.

This process was repeated after the Mannesmann deal in 2000, with a view to the creation of a new all-embracing culture – a complicated task with over 20 different networks in as many countries. In the UK, Australia, Fiji and New Zealand the company was known as Vodafone. Elsewhere, it was anything but – Misrfone in Egypt, D2 Privat in Germany, Panafon in Greece, Omnitel in Italy, Telecell in Malta, Libertel in the Netherlands, Telecel in Portugal and Europolitan in Sweden. And those are just the fully controlled subsidiaries – the associated companies included SFR (France), Safaricom (Kenya), Polkomtel (Poland), MobiFon (Romania) and Vodacom (South Africa). J-Phone and Eircell would soon be added to the list.

All these entities, which used different colours, logos and typefaces, needed to be brought together under a common name. This was done in two stages, starting with a double-barrelled new name, so that Panafon became Panafon Vodafone, Omnitel became Omnitel Vodafone, and so on. Only after consumers had become acclimatised to the new append-age was the initial name dropped. A new logo – the now familiar inverted comma – was adopted, along with a distinctive corporate colour.

Sponsorship played a key role in the process. Vodafone had been associ-ated with various sports for some years, including the Derby in the UK, the English cricket team and the Australian Wallabies, but these were essentially national initiatives. In 2000 Vodafone signed two new spon-sorship deals that were to raise its profile enormously. These were with the Manchester United football club and the Ferrari Formula One racing team. The latter deal immediately associated Vodafone with Italy's most famous racing cars and Ferrari's charismatic German star driver, six-time world champion Michael Schumacher. It was a very clever move that resonated strongly in both Italy and Germany, the former Mannesmann markets, helping to soothe some of the lingering resentment associated with the hostile takeover.

During the course of 2001 Vodafone discovered another way to project its presence on to the global stage. The acquisition of the stake in Swisscom Mobile had been driven, in part, by the company's aim of creating a single international footprint. Vodafone had a far better presence in Europe than any other telecoms company, with controlled networks in ten countries and a presence through its associate companies in another four, but the footprint was still frustratingly incomplete. The fall-out from the dot.com crash had seen its share price drop from 400p to below 100p, at which level it was no longer possible to make meaningful acquisitions without suffer-ing some degree of earnings dilution.

It occurred to Chris Gent that the high profile afforded to Vodafone by two huge deals – both associated with audacity, judgement and, most of all, success – could be exploited. The first, in December 2001, was with Tele Danmark A/S (TDC), the Danish PTT, which announced that it would be taking a licence from Vodafone and would subsequently be co-branding its mobile services. TDC's licence would give it access to all of Vodafone's new services, while Vodafone customers travelling in Denmark would find that TDC provided the same services they used at home. This would encourage them to roam on TDC's network rather than those of any of its competitors, while roaming TDC customers would gravitate towards the local Vodafone networks.

The elegance of this bilateral arrangement soon became obvious to others. A couple of months later the second deal was signed, with Radiolinja (now Elisa Corporation), the second-largest mobile operator in Finland. The Finns were obviously happy with the arrangement, as the following month they extended the deal to cover Radiolinja Eesti, their mobile subsidiary in Estonia. The gaps in Vodafone's European footprint were rapidly being filled.

By the end of the 2003 financial year, Vodafone's 'partner markets' programme had expanded to include 13 countries, including Austria, Croatia, Cyprus, Iceland, Lithuania, Luxembourg and Slovenia.

The industry understood the significance of these moves: all of a sudden, a giant competitor was providing unified services across an ever-expanding range of countries. Telia of Sweden was the first to react, announcing its merger with Sonera to create a Scandinavian champion; matters of corporate culture and any lingering national antipathy were brushed aside in the rush to increase scope and scale.

Chris Gent announced that he would be retiring from Vodafone after the 2003 annual general meeting, and used his last few months at the company to tie up a few loose ends. In late November 2002 Vodafone bought a further 7.6% of Libertel, of which it already owned 70%. A few days later, it bought out France Telecom's remaining stake in Greece's Panafon. This was followed in January by the acquisition of the remaining minority in Airtel in Spain and also a stake in Vodafone Hungary. In March it launched tender offers for all the shares it didn't own in its Portuguese, Swedish and Dutch subsidiaries. Finally, in May, it bought out a 5% minority stake in Vodafone Australia.

On 30 July 2003 Chris Gent said a heart-felt farewell to the shareholders, and all rose to applaud him. A standing ovation is a real rarity at an annual general meeting, and it demonstrated that the shareholders recognised how lucky they'd been to have had such a talented man to run their business. He would be a hard act to follow.

CHAPTER 13

Transition

Arun Sarin became Vodafone's new chief executive. News of the appointment divided opinion. Some felt that Sarin's more conventional approach after the frantic activity of the preceding four years would benefit the business; others weren't so certain. Many thought Julian Horn-Smith would have been the better choice – he'd been with the company since its inception and had a profound understanding of the culture and what it was that distinguished Vodafone from every other phone company.

Those who'd hoped that Sarin would focus solely on Vodafone's existing businesses and not go chasing further expansionist deals were soon disappointed. In January 2004, after a disappointing fourth quarter during which it recorded an $84-million loss, AT&T Wireless put itself up for sale. Cingular Wireless, the SBC-BellSouth joint venture, offered $11 a share in cash. If successful, the merger would create a new market leader, overtaking Verizon. The offer valued the equity at just under $30 billion and the business at around €36 billion, including debt.

For a day or two it seemed that this might be enough but then rumours started circulating. Vodafone, it seemed, was interested in making a bid, even though it already owned a 45% stake in Verizon Wireless. And three days later it did just that, offering $35 billion for the equity.

Barely 24 hours later, while Vodafone and the rest of Europe were still asleep, Cingular raised its offer to $15 a share, a valuation of about $40 billion, which was immediately accepted. Vodafone woke to the news that it had lost.

The contrast between the deals Chris Gent had engineered for Vodafone and this one, the first major move by his successor, was clear for all to see. Sarin hadn't done the necessary groundwork – he hadn't talked to his investors, he hadn't really explained the logic of the move and he hadn't done anything to allay inevitable concerns at Verizon – and he hadn't stayed up all night to make sure nothing untoward was happening to his deal back in his native country.

While Sarin was chasing assets in America, Julian Horn-Smith, who was close to retiring, was thinking about perhaps pulling one final rabbit out of the hat. Vodafone had acquired interests in both Poland and Romania through the AirTouch deal and these were proving quite rewarding. It didn't look as though it would be possible to take control in Poland, but Romania was promising.

During 1996 AirTouch had taken a 10% stake in a consortium called MobiFon, which was awarded a GSM licence in Romania in December that year and went on to set a new record, opening for business just 135 days after it received the licence. The company was led by Canadian outfit TIW, which held 62% of the equity. TIW was always open to offers, and had in fact sold Vodafone a further 10% in MobiFon in November 1999.

TIW's holding in MobiFon and its interest in a new operator in the Czech Republic, Oskar Mobile, were brought together under the umbrella of a new holding company, ClearWave. By 2004 the Romanian business was doing very well and Oskar was starting to pull its weight too, but Horn-Smith thought it possible that TIW might be willing to part with one or both of these assets – if the price were right. He arranged a meeting with TIW's chief executive, Bruno Ducharme, and popped both a cork and the question. Over a couple of glasses of fizzy French wine, the two men agreed to the deal, sealing it with a hand-shake. Vodafone would pay TIW £1.9 billion in cash for the 80% of Romania it didn't own, plus 100% of Oskar. The deal was completed in May 2005.

Six months later Vodafone announced that it had agreed to buy Telsim, one of Turkey's three mobile operators, in a deal negotiated and super-vised by Horn-Smith: it had been by no means straightforward. Telsim's previous owners had been sued by Motorola and Nokia in 2003, the two companies alleging that they'd been the victims of a massive – $2.7 billion – fraud. There were also issues to do with corporation tax payments – or the lack of them – so the business had been confiscated by the courts and placed in the hands of the Turkish Savings and Deposit Insurance Fund, a special-purpose body controlled by the state. Vodafone found out that the business was going to be auctioned publicly, and Horn-Smith immediately began to do his homework. What did the politicians think? What about the existing operators? What were others likely to bid? His meticulous preparation paid off, and Vodafone won with a bid at $4.7 billion. Today Telsim is highly profitable and has been for the last six years.

Shortly after this, rumours of boardroom disagreements began to surface. By the end of the 2006 financial year, seven of the nine executive directors who'd served in Chris Gent's last year were gone. A year later, only one remained – Arun Sarin himself.

During 2005 and 2006 Vodafone's new direction became clearer.

First, in October 2005, the company announced that it planned to sell its Swedish subsidiary to Telenor, a decision justified on the grounds of onerous licence conditions and severe competition.

Sarin's next move was far more controversial: Vodafone's Japanese business wasn't making the kind of returns expected, and Sarin felt it was just too difficult a market for him to crack, so he put the business up for sale in 2006. At a stroke, all the hard work Vodafone had done to obtain control of one of Japan's three national operators was undone.

The disposal was completed in April 2006, which the kinder critics have noted was 'too early'. The business was bought by SoftBank, a company that at the time wasn't especially well known outside Japan. That is no longer the case today.

Founded in 1981 by Masayoshi Son, SoftBank began as a distributor of personal computer (PC) software but with the goal of becoming 'the world's leading infrastructure provider in the information industry'.[62] The

---

62 The message is constant but the wording has been modified over the years as technology changes. This version comes from the 2004 Annual Report to Shareholders.

company grew rapidly, riding the crest of each successive wave in the IT revolution, from software to the internet, e-commerce and eventually telecommunications.

In the 1990s, before the internet was a big deal, Son founded Yahoo! Japan in conjunction with the American company. In 2003, Son had another moment of brilliance when SoftBank became a 27.6% shareholder in a new Chinese e-commerce business, the Alibaba.com Corporation, now one of the world's biggest online vendors.

Just over a year later SoftBank acquired Japan Telecom, the fixed-line business Vodafone had sold nine months earlier. This took SoftBank into the big league, and within a remarkably short time, the mobile business was producing the kind of results Vodafone had wanted but never managed to generate.

By 2008 SoftBank Mobile had taken the lead in terms of net new connections, with 47.7% of the total. Churn had dropped, the customer mix had improved, and network performance was much improved; and so, naturally, was the profitability of the business.

Vodafone had received £8.6 billion for its interest in Japan Telecom. Arun Sarin had sold a diamond – and he used the proceeds to buy a lump of coal.

Even though it had been persuaded to join the RPG consortium to bid for an early mobile licence in Madras, Vodafone had always had reservations about the Indian mobile market and, in one of his last moves, Chris Gent had disposed of the RPG stake. Sarin now looked around for another opportunity on the subcontinent, and it didn't take him long to find one. Bharti Airtel was the largest and most successful mobile operator in India.

The investment house Warburg Pincus had been an early investor but it had been slowly reducing its stake and now, as 2005 drew to a close, it was time for a final exit. It began looking for a buyer for its remaining 5.61% stake and alighted upon Vodafone, which bought it.

Sarin soon managed to raise that stake to a 10% economic interest overall – but of course, it had no say in the way the business was run. More to the point, there was already an industry partner, in the shape of Singapore Telecom, which had a 15.61% stake. It became clear that the Bharti investment was an economic cul-de-sac: Vodafone eventually concluded it had no hope of obtaining control and not even much hope of significant influence, and began looking for an alternative.

Some companies make more money from buying and selling businesses than they ever make from running those same ventures. Hutchison Telecommunications seems to be one such. During the six years it owned Orange, it recorded cumulative losses of over £1 billion, yet still managed to sell the company for £13.5 billion. It repeated the trick with VoiceStream in the USA, turning a pretty insignificant loss-making start-up into a $50-billion pile of cash.

Hutchison Telecommunications had been an early investor in the Indian mobile market and by 2006 was close to establishing a national presence. Yet all was not well within the business. Hutchison and its partner, the Indian Essar Group, disagreed on a number of key issues, not least whether to seek a listing for the business. Hutchison concluded it was time to get out and put its 67% stake up for sale.

There were several interested parties, but none quite as keen as Arun Sarin. Born into a wealthy patrician family and having lived for the first 20 years of his life in India, he felt he had a better understanding of the culture and was more at ease with the risks. 'I knew the country,' he told the *Financial Times* in 2007. 'I knew how the system worked, I knew the explosion economically that was occurring in India, and therefore could be a good advocate within the firm to say, "Fellas, we need to be here, we need to be here in size ..."'[63]

In May 2007 Vodafone announced it had secured a 51.6% direct interest in Hutchison Essar and call options that would allow it to take its stake up to 67%. The cost of this stake was $10.9 billion, some part of which it recouped immediately: the following day, Vodafone announced that a subsidiary of the Bharti Group had agreed to pay $1.8 billion for the 5.6% stake Vodafone owned in Bharti Airtel. Vodafone offloaded the balance in May 2015. Vodafone eventually managed to sell the balance in May 2015, booking a net profit of $500 million, but on balance, its Indian investments have brought it little joy and much legal wrangling, including vast claims by the

---

63 *Financial Times* interview with Arun Sarin, Vodafone chief executive, by Andrew Edgecliffe Johnson and Andrew Parker, 18 November 2007.

Indian tax authorities that continue to this day.

Vodafone has also had occasion to reconsider the value of the assets it acquired in India on more than one occasion. The doubts began to creep in during 2009, and the 2010 financial-year accounts include a £2,300-million impairment charge; further reductions feature in the accounts of financial years 2014 (£1,135 million), 2015 (£1,282 million), 2016 (£1,331 million), 2017 (€4,515 million) and 2019 (€255 million). In total, these charges have written down the book value of Vodafone's investment in the subcontinent by £11.4 billion (around $15 billion).

Do the fellas still believe they really need to be there in size, you might wonder. Arun Sarin was no longer around to ask the question, having announced his decision to retire from the business in July 2008.

CHAPTER 14

The arrival of the mobile internet

I t had taken 23 years from the first network launch to connect the billionth subscriber. The second billion came rather more quickly, just three and a half years later, in early September 2005. By this stage, the old analogue networks had almost disappeared (there were fewer than seven million such connections). The European GSM standard had become the global standard, with nearly 1.6 billion of the 2-billion total. 3G networks were only just beginning to make a mark: together, W-CDMA and the American CDMA-2000 1X-EVDO standards had a combined base of fewer than 60 million users.

Less than two years later there were over three billion connections, and just 17 months after that, in early February 2009, four billion.

At the same time, the imbalances in the market seen in earlier years were beginning to disappear – overall penetration in the world's emerging markets was just 50%, not far short of the global average of 60%. However, 80% of all mobile customers still used GSM. Nine years after the ruinous European spectrum auctions, there were still fewer than 500 million 3G subscribers.

As we saw earlier, GSM didn't really get going until Nokia produced a handset that was good enough to beat the best analogue handsets. The same thing happened with 3G – the market for mobile data didn't really take off until someone had produced a device that would allow users to take full advantage of the new services. Ironically, when that device came, it didn't even offer 3G.

The advent of Apple's iPhone changed the way consumers used their mobile handset. And that, in turn, changed the whole mobile market, shifting the balance of power away from the operators towards the equipment manufacturers and, later, the application providers. Value that had previously fallen to the networks would increasingly be captured by other businesses.

Before that though, 3G was starting to have an impact on another part of the market. In the absence of any exciting new handsets to deliver exciting new services, the mobile industry hit on a brilliant new strategy – cut the cost of voice calls! Attract traffic which had previously always been carried on fixed networks onto mobile. Why not? The limits on mobile networks' capacity had prevented this tactic in earlier years, but this was no longer an issue – after the auctions, most of the industry now had more capacity than it needed; at least, until demand for mobile data kicked in.

Slowly at first and then at an accelerating pace, phone calls that had been made from fixed lines began to be initiated on mobile networks. Tariffs that encouraged this began to emerge, as operators offered bundled minutes, 'on-net' discounts and other such blandishments to their mobile users. As mobile handsets became more commonplace, people started to realise that calling a person directly was a better option than calling a building, in the hope that the person they wanted to talk to was in that building, and close to their phone at the time of the call. The spread of mobile packages offering voicemail encouraged this phenomenon – why leave a message with an over-worked, absent-minded associate when you could leave the same message on the intended recipient's own phone? Obvious now, but less so back then.

Minute migration from fixed to mobile wasn't something the fixed-line companies had factored into their long-term planning. Two decades after the first mobile networks had been launched, many PTTs still considered the new service to be a distraction from the real business of telephony – the long-established fixed line with which they'd grown up. This was despite a rapidly accumulating body of evidence to the contrary, not to mention the fact that mobile networks were by now far cheaper to build and maintain than their fixed counterparts.

And, of course, the PTTs suffered from a fundamental structural problem: almost all of them operated both fixed and mobile networks in their home market, so they were at one and the same time attacker and defender. If the people who ran the mobile subsidiary wanted to introduce a tariff to counter some new mobile tariff from some foreign-owned new entrant, they were invariably prevented from doing so by the people who ran the fixed-line network. The fixed-line people always had the upper hand – not only was their business larger, but they invariably had better connections to the board of directors, many of whom had come up through the ranks when the fixed-line business was all there was to a telecoms company.

Up until the arrival of digital mobile phones and, in particular, the GSM system, mobile phones provided just one service – voice communications. For many users, that was enough. But digital phones could offer rather more. Voicemail was one option, caller identification another. Then there were call barring, call forwarding, three-way or conference calling … everything an advanced office PABX offered in the palm of your hand.

And then GSM came up with something PABXs couldn't do. Vodafone's 1995 Annual Report announced the launch of 'a new range of two way data services for subscribers to Vodafone's digital network including Telenote [a short-message service] which allows alpha numeric messages of up to 160 characters to be sent to and from GSM digital mobile phones.'

Texting opened up a whole new world. As Telia noted in its 1999 Annual Report, 'The service enjoyed a roaring success as it was enthusiastically received, particularly by teenagers. SMS accustoms users to new mobile techniques and acts as a gateway to more advanced services.'

By the end of the decade, operators found that they had a vast new revenue stream. And better than that, it was a revenue stream that had almost no cost attached to it – the service was (and is) carried over the GSM network's signalling channel, so it was almost entirely free to carry, and therefore the revenue from texting was almost pure profit. The 160-character limitation wasn't a problem either, as users soon developed a vernacular all their own that minimised the length of the txt. Gr8!

Those more advanced services began to proliferate and included, according to Deutsche Telekom, 'a number of programs that provide information, such as news, stock quotes, exchange rates, travel information or schedules for entertainment events'.[64]

The arrival of genuine smartphones in the first decade of the 21st century took this one step further and made the internet – once the preserve of a few bearded academics – ubiquitous. Anywhere and everywhere, people can access the net through a handheld device.

As with almost every other development, there are arguments about who produced the first smartphone; this is one version.

The first handheld device that came close to providing mobile internet access was the Psion Series 3, which was launched in 1991 by British company Psion plc. The clam-shell Psion opened to reveal a full 58-key QWERTY keyboard and a monochrome display. It used an operating system called EPOC, which supported a simple but powerful suite of applications, including a database, word-processing and spreadsheet functions, and other standard applications such as a calendar, world clock, address book and calculator. It came with removable flash memory drives, used standard AA batteries and could be connected to the internet through an optional external modem.

The concept was soon emulated by other manufacturers, and in 1992 Apple launched its Newton, the first personal digital assistant (PDA). In 1994 IBM came up with Simon, the first PDA to offer mobile connectivity.

For the game changer, however, we had to wait until 2001, when Nokia launched the third in its 9000-series communicators, the 9210. For this new model, Nokia abandoned its home-grown operating system and switched to new software called Symbian, a development of Psion's EPOC. This gave the 9210 the functionality of the new Psion 5-series organiser, plus mobile connectivity over GSM and, best of all, a full-colour screen. It was a remarkable device, but its high cost – around $1,500 at the time of its launch – put it beyond the reach of the vast majority of the market.

Eventually, other, cheaper, PDAs from companies like Palm and

---

64 Deutsche Telekom, 2000 20-F filed with the American SEC.

Compaq came on to the market and Psion's glory days were over. It abandoned its consumer businesses to focus on rugged mobile devices and was later acquired by Motorola. But by then, Psion had helped establish a benchmark for all PDAs and the smartphones that were to follow them. Mobile-handset vendors now had a target to aim at: the market wanted powerful devices with a lot of inbuilt functionality but that were at the same time lightweight, intuitively easy to use and, very importantly, attractive in appearance.

Apple got the point, but curiously, almost none of the other computer manufacturers did. Although IBM had helped start the trend towards smartphones, it was nowhere to be seen. Compaq and Hewlett Packard had both enjoyed some success in the PDA market but neither of them managed to replicate this in the telephone market. In many ways, this failure is quite extraordinary, as these and other similar companies had spent the previous 30 years telling their shareholders of the golden age that was to come once the computer and communications industries had 'converged'.

Apple's great triumph was that it managed to integrate most of the functions of a mobile phone with those of a PDA in a single unit. The new iPhone featured a touchscreen and a camera, plus a battery capable of supporting eight hours of talk time. The battery life apart, that may look like a pretty lacklustre spec by modern standards, but to appreciate what a revolutionary – if imperfect – device the iPhone was, we need to look back at the state of the handset market in the last years of the 20th century and the first years of the 21st.

The rising number of new mobile customers obviously represented a huge opportunity for the mobile-handset industry. In 1997, the market shipped over 100 million units for the first time, thanks to the happy combination of new users (68 million in that year) and a growing replacement market, as existing subscribers either traded up or migrated to digital devices. Nokia, whose launch of the 2110 in 1994 had helped establish the company as a leading supplier of GSM phones, thrived in this environment.

Nokia had gone all-out to perfect devices based on the European standard, which it had had a hand in developing. The 2110 was followed by a succession of elegant, desirable handsets (and the occasional flop).

Crucially, Nokia recognised the importance of continuity, and all of these devices used the same menu structure, which was logical and intuitive. If you'd mastered the 2110, getting to know the 6110 or the 8110 was the work of a matter of moments.

However, by the middle of the new century's first decade, there were a few warning signs. Nokia had abandoned the simplicity of its earlier interface and the N95 in particular had a labyrinthine menu. The company's high-end devices weren't receiving the rave reviews they had in years gone by; and the average selling price was slipping, inexorably, year by year. But such was the momentum of the business that these red flags were either ignored or perhaps not even noticed. The company was routinely supplying 35–40% of the world's handsets, more than its two largest competitors, Motorola and Samsung, put together.

The arrival of the iPhone opened the eyes of consumers all around the world, if not those in the boardroom at the Nokia headquarters in Finland. Suddenly, here was a device that offered something you just couldn't get from a Nokia N95. No, it wasn't the brilliant colour display, the 5MP camera, the GPS-based maps program, the music player or even the 3G+ high-speed downlink packet access (HSPDA) capability. It was simplicity.

The iPhone was immediately intuitive, while Nokia had finally abandoned the straightforwardness of its earlier interface. Comparing the iPhone's specifications to those of the N95, the latter was better on all counts apart from the one that really mattered.

When Apple launched a 3G version of the iPhone in June 2008, the Finns had to finally sit up and take notice. In a single quarter, Apple sold nearly 7 million units – with a price tag some eight times higher than Nokia's average.

The American company had adopted a rather curious approach to marketing its new mobile device. It selected a single operator in each country as its chosen distributor for that territory – AT&T in the US, for example, O2 in the UK, T-Mobile in Germany and Orange in France. These companies paid for the right to sell Apple products to their customers, and there wasn't even much of a protest when the chosen carriers discovered that the iPhone didn't receive signals quite as effectively as the Nokias it was beginning to replace. Radio-frequency (RF) engineers muttered to each other as they went about re-engineering their networks to cope with the device's foibles.

At the time Apple launched its first 3G version of the iPhone, there were some 3.78 billion mobile connections worldwide. Of these, fewer than 300 million used the W-CDMA standard while only about 10% were capable of accessing a 3G network of any kind. A year later, at the end of 2009, 3G phones accounted for about 30% of all net additions, while the total in use had doubled. And by the beginning of 2011 there were more than a billion 3G phones in the world and Apple had sold the best part of 125 million iPhones, the vast majority of which were based on one or other 3G standard. Remarkably, its average selling price was almost unchanged, at around $650 – about $300 more than the second-best in the industry, Taiwan's HTC.

Smartphones rapidly expanded the capabilities of the mobile phone beyond voice and text messaging. Users could store the details all of their contacts, see their daily schedule on a calendar, play games, listen to music and take photographs. Email, instant messaging, and the ability to watch TV or access the internet were to follow.

An entirely new industry grew up, developing customised applications – apps – that could be downloaded on to the device. Users now had access to word-processing software, spreadsheets and databases. The camera – or cameras – could be turned into document scanners. The phone's microphone was capable of recording, while voice-recognition software gave users another way to access the internet. Location data from the GPS system could be displayed on moving maps, while a synchronised, synthesised voice provided navigation assistance.

This brief list merely scratches the surface. The range and variety of apps that are currently available is little short of astonishing – a world of possibilities available at the click of a button through an elegant metal box at a cost of a few hundred dollars.

And the capabilities of these devices just keep on improving, at a rate that seems to be accelerating. Processors keep getting faster, displays are brighter, denser and clearer, there's more memory and far more storage, and the cameras offer resolutions that put most digital SLRs to shame. Dual-band phones that could work on both 900MHz and 1800MHz networks were once the big thing; today's top-end phones are all multi-band, offering access to 20 or more frequency bands. It's no longer just landlines that are under threat, it's also cameras and video recorders, hifi, computers, SatNav devices, television, and even the mobile networks themselves.

In the beginning, the most expensive aspect of mobile communications was the acquisition of the phone itself, with prices of more than $1,000 not unusual. Over time that $1,000 became $100 or less, and the value in the industry shifted to network access and call traffic. Average users might generate monthly revenues of $40–50, while a business user could easily run up a monthly bill of five or even ten times that. The smartphone changed that back again, and today, mobile-phone users in the developed world can access the internet all the time and make all the calls they need for $20 a month or less. If they travel a lot, they might be spending $50. That's less than one twentieth of the cost of a new top-end iPhone X.

Value within the telecoms industry – and especially the mobile side of the industry – has drifted away from the telecoms operators towards companies like Apple, Amazon, Facebook and Google, companies that would be out of business or perhaps would never have come about in the first place without access to telecommunications. The mobile operators have been heavily criticised by some for allowing this to happen, but you have to ask, what could they have done to prevent it? They provide the infrastructure for today's knowledge economy; they are the modern-day counterparts of the roads and railways of the 20th century. Commerce needs roads to provide access to customers, just as Facebook needs AT&T and Vodafone, but no one could reasonably suggest that the value of our roads is greater than the value of the goods transported across them. That's why it's called *infra*-structure (from the Latin for 'beneath'): networks of this kind come below the traffic they carry in the value chain.

Apple had taken a huge bite out of the handset market. It had seen off Nokia, its largest and most powerful competitor, which had realised it couldn't compete in this changed world. It sold its handset business in 2008 to Microsoft.

Yet Apple wasn't destined to have the market all to itself. In 2005 a new competitor emerged from a rather unexpected direction when American technology company Google bought a small software start-up called Android. Android was a free open-source mobile software platform that allowed developers to create applications for mobile devices. 'Android is being developed with the Open Handset Alliance, a group of more than 30 technology and mobile companies, with the goal of providing consumers a less expensive and richer mobile experience', Google noted.[65]

Google launched the technology in September 2006, giving the world 'a new mobile operating system that would allow open interoperation across carriers and manufacturers'. 'Anyone is free to use it and modify it', they said.

With over 20,000 applications, Android made possible for users an 'open, internet-enabled computer in their pocket that is as good as a laptop from a couple of years ago, [which] has no trouble playing music ... over their car stereo, interrupting to read street names and display a map ... driving directions that prompt you, just like a real navigation system ... updated traffic and even a photo ... of your destination.'[66]

In mid-2019 we saw exactly how powerful Android had become when a trade dispute between the US and China led to President Donald Trump's banning the use of any equipment made by the Chinese firm Huawei in any US network, and Google fell into line with this command, withdrawing support for its Android operating system on all Huawei smartphones. Huawei's vast user base could lose access to key Android features such as Gmail, Google Maps and the Play Store, greatly diminishing the utility of the smartphone. Since Huawei is now the third-largest manufacturer of these devices, after Samsung and Apple, the implications are profound.

---

65 Google 2007 10-K.
66 Larry Page and Sergey Brin, 'Founders' Letter', Google 2008 10-K.

In April 2002 the mobile industry connected its billionth subscriber. Four months later, mobile connections outnumbered fixed for the first time. Even the most traditional phone companies now realised that the mobile phenomenon was not a flash in the pan: it was here to stay.

Customers began cancelling their fixed-line phone contracts. The reduction in lines was barely noticeable at first. It began in the US, as families started eliminating the second (or even third) line in the house, which had been rendered superfluous by growing mobile ownership. Equally, businesses began to consider whether they needed quite so many fixed lines, given that a growing number of employees now used company-sponsored mobiles. Lines that had been dedicated to fax machines were also beginning to be disconnected, as email started to supersede the older technology.

The cable companies were also starting to make an impact. As these networks were generally capable of delivering higher-speed data than traditional copper fixed lines, a growing number of consumers started using them for internet access, rather than relying on dial-up connections.

Phone companies were in an unenviable position: they had to either accept the demise of the traditional network and manage the decline to the best of their ability or face the enormous cost of rebuilding the entire network with modern fibre-optic components. Neither option was remotely attractive.

Fortunately, a third materialised. Technology was working to improve mobile services, but it could also be used to improve the fixed network. Bell Labs had begun work on a concept called Digital Subscriber Line (DSL) in the early 1980s, which, it was hoped, might eliminate or reduce the problems associated with traditional copper networks, such as low data capacity and high noise. By 1989 it had managed to develop the concept to the point where it could be utilised commercially.[67]

DSL divides the bandwidth on a copper pair of lines into three unequal parts, one of which carries the traditional voice signal while the other two are used for data. In the most common variant of the technology, Asymmetric DSL (ADSL), the available bandwidth is split unevenly in

---

[67] By this time the work had passed to Bellcore, a research body that was jointly owned by the seven regionals.

favour of the download path. The speed at which data can be transmitted over such a connection is typically about 6Mbps but it is determined by several factors, including the distance between the exchange and the subscriber's premises, the age and quality of the copper, and the whim of the network operator. (There are now several versions of DSL, including VDSL or Very High-Speed DSL, which can allow transmissions of 50Mbps or even more over short distances.)

Deutsche Telekom was a very early adopter, beginning to enhance its local loop (the connections between the local exchange and its customers) with the ADSL variant in 1996. BT followed soon after, with several of the Bell companies and France Telecom following suit in 1998, after which it became increasingly widely utilised.

DSL offered fixed-line operators a way of enhancing the speed and overall utility of their networks without the massive cost of replacing copper with fibre. However, it's generally seen as something of a stop-gap ahead of the wholesale deployment of fibre, which is invariably referred to as 'future proof' by its proponents.[68]

For now, though, future proof has to give way to current cost: a network of this kind is unaffordable – not so much because of the cost of the components as the cost of construction. Fibre is delicate and doesn't easily bend – it's glass, after all – and it needs to be protected every inch of the way. Roads and other thoroughfares need to be dug up and expensive optical splitters need to be connected to each individual building. That all costs money.

In the meantime, mobile technology is also improving. The first 4G network was brought into service in Europe at the end of 2009, and today it's the industry's dominant technology, accounting for nearly 4 billion of the world's total mobile connections. Although not often achieved in practice, 4G networks can theoretically support download speeds of 300Mbps, or just under two minutes of ultra-high-definition video every second. Not

---

68 But is it really future proof? The fibre used in the first fibre-optic submarine cable, TAT-8, brought into service in 1988, was capable of carrying 280Mbps; three decades later, the latest fibre technology is capable of delivering 1Pbps – 3.8 million times more, and the rough equivalent of 10,000 hours of ultra-high-definition video every second. In what sense, then, was TAT-8's fibre proof against the future?

many fixed connections achieve this speed.

And now the first 5G networks are being deployed. In the laboratory, such networks have achieved speeds of 3Gbps, or ten times the speed of 4G, while the first commercial networks in the US are currently achieving speeds of over 1.3Gbps. Ultimately, 5G is expected to reach 10Gbps, way faster than the vast majority of fixed fibre connections.

CHAPTER 15

Consolidation, convergence and the struggle for value

As the 20th century came to an end, American mobile operators were still some way away from achieving full, national coverage over a single, unified network.

The 110 million mobile customers in the country at the end of 2000 were still divided between numerous different operators, very unevenly. Four large operators dominated the market, with a combined market share of 67%; the next three shared a further 16%; and the 17% balance came from the many, many smaller regional operators. Of these, only one had more than a million customers.

At the head of the list were two entities: Verizon, the larger of the two, created by the merger of Bell Atlantic and AirTouch and a subsequent merger with GTE; and Cingular, the company that had emerged after the mobile units of SBC and BellSouth had come together. The one-time market leader, AT&T Wireless, was third, ahead of Sprint.

By 2003 the picture was similar – Verizon still led, with 37.5 million of the 157 million total, Cingular and AT&T Wireless were still in second and third place, and still ahead of Sprint – the most noticeable difference was that there was now another major player in the shape of Deutsche Telekom's T-Mobile US, which now had over 15 million customers, having nearly tripled its base over the three years. The big four had become the big five.

At the start of the new century, AT&T had appeared well positioned to face the challenges of a changing market. The deals done in the late 1990s with TCI and MediaOne (once a subsidiary of US West, focused on mobile and cable TV) had made it by far the largest cable-TV company in the country, operating under the name AT&T Broadband. It was still the largest company in the long-distance market, even if that activity no longer provided a guarantee of prosperity in the way that it once had. Through AT&T Wireless, it was a respectable – if not especially profitable – number three in the mobile market and of course, it was still one of the world's largest business services companies.

Its position in this fast-growing marked had been underpinned in early January 2000 by the acquisition of MCI's interest in the Concert joint venture with BT, which was designed to be 'a leading communications provider for multinational business customers, international carriers and ISPs worldwide'.[69] Even if the overall level of profitability wasn't all that it might be, the company looked – at least to outsiders – to have emerged from a period of dramatic change in reasonable shape.

Appearances are often deceptive. By the end of the following year, there wasn't a lot left of the business that had been the world number one for so long. In May 2001, AT&T announced that it was creating a separate 'tracking stock' for AT&T Wireless; by July, the mobile company had been spun off as an independent business, its parent retaining just 7.3% of the shares. It disposed of those in the last weeks of the year. A month later, AT&T spun off Liberty Media, a subsidiary of TCI that had specialised in TV-programme production and content. In December, it announced that it intended to sell AT&T Broadband to Comcast, creating a new market leader in cable, which AT&T referred to as AT&T Comcast. The Philadelphia-based cable operator had other ideas – it dropped the AT&T part of the name immediately after completing the deal.

Stripped of mobile, its cable-TV business, the manufacturing arm and the seven regional companies, the 21st century version of AT&T was a sad shadow of its former self. It limped along for four more years until one of its children decided to put it out of its misery. In November 2005, SBC bid for AT&T, buying the business for a mere $18 billion. For that modest

---

69 British Telecom 2001 20-F.

outlay, SBC got control of the long-distance business and of course, the name – that wonderful, resonant name that had adorned the company since 1881. If Comcast hadn't wanted it, SBC certainly did – it dropped 'SBC' like a hot potato and emerged as AT&T. The rebranding was obviously the right move, swapping a locally famous name for one of the few genuinely iconic names in the whole of industry, but it has caused a certain amount of confusion among the general public ever since.

The 'new' AT&T (based in Texas, not New York) now owned three of the seven regional Bell companies, plus its former parent, the long-distance company, and a 60% share of the largest mobile business in the US. In 2006, it acquired yet another of its sister companies, its southern neighbour, BellSouth, giving it 100% ownership of Cingular, plus 20 million more access lines which generated annual revenues of over $20 billion. The deal was valued at $86 billion. Although the antitrust authorities were spitting blood, the competition commission knew there was no case to answer – the two operated their regulated telephone businesses in contiguous territories and didn't compete, so the acquisition couldn't be deemed anti-competitive.

Four of the original seven Bell regionals were now back under the one roof, together with the nation's largest long-distance company. But the enlarged AT&T knew it would have to fight hard if it were to fend off attacks from companies like Comcast, the giant cable company its former parent had helped to create. Minute migration had affected fixed revenues and the cable companies' entry into the local phone market had exacerbated that problem. The fight for internet subscribers and entertainment customers wasn't always a fair one: many of the newer cable systems were technically superior to the Bell System's traditional copper networks and were able to offer packages that combined television, internet and voice services.

Verizon was feeling the same kind of pressures. Verizon Wireless (the company that had been created by the merger of its mobile business with Vodafone AirTouch) was going well, at least as far as day-to-day trading was concerned, but there were problems with the 55%/45% ownership arrangement. Although it was a smaller business than Verizon's traditional fixed-line one, Verizon Wireless was far more profitable, generating more than half of group profits and most of its available cash flow. Verizon now wanted to be rid of its British partner and had stopped the wireless subsidiary from paying dividends, in the hope that this would focus minds back at Newbury. However, Vodafone had no pressing need for the cash that would bring and so was in no hurry to sell.

As time went on, the decision to eliminate the dividends at Verizon Wireless began hurting the whole Verizon Communications group. Although the consolidated balance sheet looked all right, without dividends from Wireless, disposable cash was hard to find. It needed to enhance the profitability of its fixed business, which was investing heavily in its new fibre network at the time. So it looked to strengthen its position in the market by buying MCI. The irony in the move wasn't lost on anyone: MCI, the principal agent in the dismantling of the Bell System 25 years earlier, was being bought by one part of the thing it had fought so hard to destroy.

At the same time as Verizon was focusing its gaze on MCI, two of its competitors in the mobile market, Sprint PCS and Nextel, had announced their intention to join forces. The two companies were really rather different, with different technologies – CDMA for Sprint PCS, and integrated digital enhanced network (iDEN) in Nextel's case – and very different business philosophies, but, as the recent spate of transactions was making abundantly clear, size mattered. The $37.8-billion merger was completed in August 2006, giving Sprint-Nextel 47.6 million mobile customers and annual revenues of some $35 billion, and creating a clear number three in the US mobile market.

All the while, the market for mobile services continued to grow strongly. At the end of 2003, there had been 157 million connections; by the end of 2005 the total had reached 208 million. AT&T still led, with 54 million, but Verizon was creeping closer, with 51 million, while Sprint-Nextel was closing in on that with 47.6 million. T-Mobile was a distant fourth with 21.7 million, and Alltel, a specialist in rural and suburban markets, accounted for another 10.7 million. The rest were scattered between countless small independent operators.

Subscriber numbers continue to build throughout 2006 and 2007, during which time AT&T bought Dobson Communications, one of the larger independents, to extend its lead over Verizon. Verizon responded by taking out Rural Cellular then, in June 2008, it edged ahead, acquiring Alltel for $28.1 billion. Verizon now had 91 million mobile customers, 11 million more than AT&T and 9 million more than the combined total of Sprint-Nextel and T-Mobile US. The three other independent businesses of any size, Leap Wireless, MetroPCS and US Cellular, between them had 17.7 million connections, just over 6% of the 286-million national total.

By 2011 Verizon had passed the 100-million mark. To get back on level terms AT&T needed the best part of 20 million additional subscribers, and to this end it announced that it had reached an agreement with Deutsche Telekom to acquire the whole of T-Mobile US at a cost of $39 billion.

The German company had been finding life increasing difficult against larger, better-positioned competitors. It had built up a base of over 33 million customers, but it was a continuous struggle. The bankers had told the German parent that they had too much money tied up in the USA, in the same way that it had had too much tied up in the UK (thanks to the acquisition spree of 1998–2000). The threat of a debt downgrade hovered somewhere in the background (as it so often did in the first decade of this century) and so an exit seemed timely. AT&T's offer gave it one – and, remarkably, it needed only a little creativity in the way the numbers were struck for the deal to generate an overall profit.

Both sides were happy – but the frustrated bureaucrats in the antitrust division of the US Department of Justice were not: this was clearly anti-competitive, they said, and the courts and the FCC agreed with them, citing likely undesirable results of the deal that included 'increased prices for consumers, reduced incentives for innovation, and decreased consumer choice'.[70]

As a consequence, and adding insult to injury, AT&T ended up paying Deutsche Telekom a $3-billion break-up fee for the failure of the deal.

---

70 Bureau Dismissal Without Prejudice of AT&T's Applications for Transfer of Control of T-Mobile USA, Inc., 29 November 2011.

On the other side of the Atlantic, Vittorio Colao, Vodafone's new CEO, took over from Arun Sarin in July 2008. The former Omnitel Pronto Italia executive was in no hurry to enter the merger-and-acquisition fray in the way his predecessor had been. He had other matters to attend to. The company was still only achieving modest earnings growth, thanks in large part to cost-cutting measures and the strong performance of Verizon Wireless, but mobile markets all around the world were also becoming increasingly competitive.

And, of course, the effects of the 2008 global financial crash didn't help. 'The economic downturn is affecting Vodafone in several ways,' Colao noted, citing a decline in voice and messaging revenue, a reduction in roaming revenue due to lower business and leisure travel, and a drop in enterprise revenue 'as business customers reduced activity and headcount'.[71]

Vodafone had backed off from Chris Gent's vision of a global presence and was now focused on Europe and a few other markets such as India and South Africa. But that wasn't really the worry – of greater concern was the feeling that the company was no longer special, no longer very different from any other telecoms operator. Having the inside track is only an advantage if you're running in the right direction around the track, and it wasn't obvious that Vodafone was.

While the company's 2009 annual report declared that it was positioning itself for changes in the industry through a 'total communications strategy to deliver broadband and internet offerings',[72] this 'positioning' was unhurried. The first moves towards the new total communications strategy had been the purchase of two small fixed-line businesses in Italy and Spain in December 2007 and the acquisition of the outstanding 26.4% minority shareholding in Mannesmann's fixed-line business in Germany six months later.

The next move didn't occur until July 2012, when Vodafone spent £1.5 billion acquiring the remnants of its one-time predator, Cable & Wireless Communications. The deal gave Vodafone access to a 20,500-kilometre fibre network (the old figure of eight) and a number of international

---

71 Vodafone Annual Report, 2009.
72 Ibid.

customers, but while it would provide cost savings on backhaul and under-pin Vodafone's planned move into the DSL market in the UK, it didn't really look like any kind of fixed/mobile integration or converged-services play.

Then, in June 2013, Vodafone announced that it had bid €7.2 billion to acquire a controlling stake in Kabel Deutschland, a publicly listed com-pany that operated the largest cable-TV network in Germany. Vodafone was no stranger to Germany, owning the second-largest mobile operator in the country and also a fixed network, but it had no presence in cable (and indeed no experience of cable anywhere). Liberty Global, a multinational cable company that owned Unitymedia, the second-largest German cable company, topped Vodafone's figure – but only just – with a €7.5-billion bid. Kabel Deutschland turned both down.

Vodafone considered its position, then settled the matter with a higher bid, valuing the business at €7.7 billion, which was accepted.

Back in the US, Sprint was in a parlous position. Then, in 2012, an offer of help came from an unexpected direction. In October it announced that it had reached an agreement with SoftBank whereby the Japanese company would invest $20.1 billion in Sprint, $8 billion of which would be used to strengthen the company's financial position.

The transaction, said SoftBank's CEO Masayoshi Son, provided an excel-lent opportunity for SoftBank to 'leverage its expertise in smartphones and next-generation high-speed networks, including LTE [long-term evolu-tion, 4G mobile technology], to drive the mobile internet revolution' in the USA. 'As we have proven in Japan, we have achieved a V-shaped earn-ings recovery in the acquired mobile business and grown dramatically by introducing differentiated products to an incumbent-led market,' he said. 'Our track record of innovation, combined with Sprint's strong brand and local leadership, provides a constructive beginning toward creating a more competitive American wireless market.'[73]

---

73 SoftBank Group press release, 17 October 2012.

But SoftBank's involvement didn't result in the same 'V-shaped' impact it had had in Japan, and the company reported huge after-tax losses in 2012, 2013 and 2014.

Son decided to double up his bet, and in December 2014 announced that Sprint would bid for T-Mobile, and a figure of $20 billion was mentioned for an unspecified majority stake. The proposal immediately attracted adverse comments: a merger of the number-three and number-four players would leave the US with only three operators of any size which, it was suggested, would damage competition. That the bankruptcy of one of the operators would have the same effect went unremarked.

By April the proposed investment had risen from $20 billion to $24 billion, but there was not much more flesh on the bones.

The months went by and then, on 1 August, Iliad SA, a small but disruptive French operator, muddied the water by bidding $15 billion for a 56% stake in T-Mobile, valuing it at $28 billion overall. This wasn't much more than a 5% premium to its market capitalisation at the time and $10 billion more than Iliad's, so no one took it seriously, but it was enough to upset Son. Three days later he announced that plans for the T-Mobile takeover had been abandoned.

He might have added 'for the moment' because Sprint's position stubbornly refused to improve. Although they were reducing, the losses continued, the $3.35 billion of the 2015 financial year fell to $2 billion in 2016 and just $1.2 billion in 2017. 'Just' – that's still $3.3 million a day. Mobile-subscriber growth had slowed, and in certain markets subscriber numbers were flat if not actually dropping. The price of mobile and fixed services was under pressure and so, inevitably, was profitability.

In September 2013 Verizon announced that it had agreed to buy a 45% stake in Verizon Wireless from its partner, Vodafone Group, for $130 billion (£80 billion). It was 230% of the amount Vodafone had paid for the whole of AirTouch Communications 15 years earlier in a deal which, at the time, some investors had criticised as ill-considered and a destroyer of shareholder value.

The deal closed in February 2014 and generated plenty of publicity – everyone, as usual, had an opinion. Some of the more astute commentators highlighted a few pertinent questions: what did this move say about the British company's strategy? Surely if it was selling out of the USA, that was tantamount to giving up? Or did it have something else in mind for the US? Eventually it became clear that Vodafone was leaving America and had no plans to re-enter the market in any way.

Attention then switched to the next big issue – what should it do with the money? Awash with cash, Vodafone elected to pay a staggering $85 billion – well over half the total – straight back to its shareholders by way of a special dividend. This set another record, being by far the largest one-off payment to shareholders in corporate history.

The company then decided to use some of the remaining proceeds to buy a second European cable business, Grupo Ono in Spain. Ono had been gearing up to go public, but Vodafone's offer of €7.2 billion was enough to persuade it not to bother – and the deal was done. Colao told investors that 'the combination of Vodafone and Ono creates a leading integrated communications provider in Spain and represents an attractive value creation opportunity for Vodafone'.[74] Perhaps. Other commentators suggested the name said it all.

For its part, Verizon's purchase of Vodafone's stake in the mobile venture had come at some cost to its balance sheet – it's net debt exceeded $100 billion. Its next moves were therefore necessarily modest, but they pointed to a new strategic direction. Between late 2013 and early 2014 it acquired the assets of three small companies as part of a plan to strengthen its position in video content and distribution: upLynx, which specialised in uploading and encoding TV; EdgeCast Networks, to help with online digital media content; and Intel Media, a business dedicated to the development of cloud TV products and services. According to Verizon, the transaction would 'accelerate the availability of next-generation video services, both integrated with Verizon FiOS fiber-optic networks and delivered "over the top" to any device.'[75]

---

74 Report in the *Financial Times*, 17 March 2014.
75 Verizon Communications Investor Quarterly, Q1 2014.

In May 2015 Verizon took another step in this direction when it acquired AOL. The one-time internet high flyer had fallen a very long way since its \$160-billion merger with Time Warner in 2000, and Verizon was able to buy the remnants of the business for just \$4.4 billion. The principal attraction was AOL's expertise in video advertising, but by the standards of the time it was a pretty low-key deal.

AT&T was also thinking about strengthening its digital entertainment business, and in July 2015 it bought DirecTV, the larger of the two DTH satellite broadcasting companies in the US. As AT&T explained, 'The nationwide reach of DirecTV and superior content-owner relationships significantly improves the economics and expands the geographic reach of our pre-existing AT&T U-verse video service'. According to AT&T, this was not just a move to protect U-verse from Amazon and Netflix.

'We believe it is critical that we continue to extend our brand leadership as the premium pay-TV provider in the marketplace by providing the best video experience both at home and on mobile devices. We believe that our flexible platform that uses a combination of satellite, IP [internet protocol]-based and cloud infrastructure with a broadband and wireless connection is the most efficient way to transport content to subscribers when and where they want it. Through this integrated approach, we're able to optimize the use of storage in the home as well as in the cloud, while also providing a seamless service for consumers across screens and locations'.[76]

'When and where they want it': patterns of behaviour were changing and the television set in the living room had lost its monopoly to laptops and smartphones. To stay ahead of the game, AT&T had realised it had to be able to deliver anything to anyone at any time, irrespective of location. At a stroke, the DirecTV deal made AT&T the largest supplier of digital entertainment services not just in the US but in the world.

AT&T looked well positioned for a converged future but it wasn't finished yet. In October 2016, little more than a year after the DirecTV deal, it

---

76 AT&T Form 10-K 2015.

announced its intention to acquire Time Warner, one of the world's largest media companies. This was an even bigger bet than DirecTV – the offer valued the business at $106.4 billion including debt. The combination, it was claimed, would create a company with the 'best content and distribution across mobile, TV [and] broadband'.[77]

The AT&T-Time Warner combination would not only be the world's largest pay-TV company and one of the world's largest telecommunications and internet businesses, it would also own a huge amount of content that it could distribute over its fixed, mobile and video networks. It would be able to charge consumers for the content and then again for the cost of transmission, while also receiving additional revenues from advertisers. Knowing who watched what – and how and where they watched it – would allow the company to 'use data and analytics to enable advertisers to better reach their target audiences and improve their ability to measure the effectiveness of their advertisements'.[78] And this, of course, would raise the price the company could charge for the use of those advertising slots.

It looked like the deal was a 'win-win-win', but once again there were protests from consumer groups, so the acquisition was scrutinised by the courts. After deliberating for nearly two years, the deal was allowed to proceed in June 2018. The Department of Justice immediately appealed, but the courts declined to reverse their decision and, for the moment, at least, there has been no further move to unravel the deal.

The merger between AT&T and Time Warner raised a number of interesting questions. It was a huge deal, with correspondingly huge risks attached. Was AT&T right in its thinking that it needed to own and create proprietary content, rather than just buy it off the shelf? Before the merger, Time Warner hadn't seemed to think so, as it had only recently disposed of its cable-TV business. None of AT&T's competitors had done anything of the kind, but then there were only a limited number of companies in the media and content business, but hundreds of potential telco distributors.

The 2017 Time Warner 10-K noted that 'the Company's businesses operate in highly competitive industries, and the environment in which the

---

77 AT&T Time Warner Analyst Call presentation, 24 October 2016.
78 Time Warner 10-K 2017.

Company competes has become more challenging over the past few years due to the shifts in consumer viewing patterns … increased competition from OTT [over the top] services and the expansion by other companies, in particular technology companies'.

But it didn't mention any of these companies by name, because in the Western world where it operated, there really was no one else in the same league. The opportunities for a similar vertical integration were few and far between, and so far, Verizon hasn't found one. Neither has anyone else.

So, is the 2018 AT&T more or less of a monopoly that the 1983 version? It doesn't control the local phone business in 26 of the 50 states, but it does provide mobile communications, satellite TV and long distance in 48 of the 50, and in addition it owns a huge amount of high-value media content, both past and present. In the meantime, Verizon has continued to build up its digital-services side, and most recently, in June 2017, it acquired Yahoo!, the one-time undisputed leader in the internet search business. Yahoo! was merged with AOL and Verizon's other media assets in a new division branded Oath.

Sprint-Nextel isn't even at the starting line in this race for content. Even with Masayoshi Son's help, it had dropped to a bad fourth place in the market. T-Mobile, meanwhile, is going from strength to strength and is on course to report a near-doubling in profits. It is, at last, beginning to justify Deutsche Telekom's decision to acquire the business.

In April 2018 Sprint-Nextel and T-Mobile announced plans to merge their businesses under the T-Mobile label. Once again, consumer groups began bleating about reduced competition, higher prices, etc. The FCC was minded to give the deal its approval, but the antitrust lawyers had other ideas. At the time of writing, the outcome of the bid remained undecided.

In the second decade of the 21st century, telecoms mergers and acquisitions seemed to have been almost exclusively the preserve of three or four giant companies. AT&T had done its deals, Verizon its, so once again it was Vodafone's turn. In May 2018 it announced that it had reached an agreement with Liberty Global to acquire the US company's cable-TV operations in Germany, the Czech Republic, Hungary and Romania for €18.4 billion (about £16 billion).

'This transaction will create the first truly converged pan-European champion of competition,' Vodafone CEO Vittorio Colao said. 'It represents a step change in Europe's transition to a gigabit society and a transformative combination for Vodafone that will generate significant value for shareholders. We are committed to accelerating and deepening investment in next-generation mobile and fixed networks, building on Vodafone's track record of ensuring that customers benefit from the choice of a strong and sustainable challenger to dominant incumbent operators. Vodafone will become Europe's leading next-generation network owner, serving the largest number of mobile customers and households across the EU.'[79]

All the buzzwords were there and it sounded very plausible, but would it really 'generate significant value for shareholders'? And if it would, why did Liberty sell it? Mike Fries, Liberty's CEO, explained, 'We are not saying what we will do with the proceeds; we have no transactions in the queue. We intend to look to optimise growth and shareholder returns. This is about us being clear-eyed about where our best opportunities lie.'[80]

'Our best opportunities': this suggested that Liberty thought that a pile of cash was better than a set of European cable-TV businesses. Vodafone disagreed to the extent that it considered these assets so attractive that it was worth disappointing its shareholders in the short term (by cutting its dividend by 40%) in order to afford the outlay.

Could they both have been right? It was quite possible for an asset to be worth more to one party than it was to another, and Vodafone was now in a position to offer quad-play packages (cable TV, internet, fixed telephony and mobile) in all five markets – Germany, the Czech Republic, Hungary, Romania and, through the Ono deal, Spain. This wasn't a bad position to be in – multi-play operators tended to generate significantly higher revenues

---

79 Vittorio Colao, quoted in the UK press.
80 Mike Fries, quoted in the UK press.

per customer than their standalone equivalents and also enjoyed far lower churn rates.

But would the market stay like this forever? Or even for very long?

Older people tend to watch television, make the occasional phone call to their friends and occasionally access the internet, generally using a PC or perhaps a tablet. Younger people still watch TV, but not nearly so much of the time, preferring to send messages or access social media and other internet sites from their smartphones. Recent surveys of consumer behaviour in the developed world suggest that these differences in behaviour and preference between the old and the young become more pronounced the further one goes down the age spectrum.

In the recent past, a number of the largest US cable-TV companies – all of which have some kind of multi-play offerings – had seen growing numbers of customers cancelling their subscriptions to entertainment packages, preferring instead to use services like Netflix or Amazon Prime; or perhaps some of the numerous illegal file-sharing services that can be found on the internet. While it would probably be reckless to extrapolate from this that in the future no one will watch TV, it doesn't need a change of that magnitude to call the investment in cable into question. Could it be that this deal – and its two forerunners – would prove little better at generating shareholder value than the ill-judged Indian investment?

Just three months after the cable acquisition, in August 2018, Vodafone announced that it intended to merge its Indian mobile business with that of IDEA Cellular, a publicly listed offshoot of the Aditya Birla Group. Both companies had been hard hit by Reliance Jio's arrival in the market and it was clear that the move was primarily defensive. Press releases talked of 'merger-related synergies' and 'annual run-rate savings' but the fact is that Jio had well and truly broken the old model, seemingly beyond repair.

Vodafone probably recognised this, and one benign effect of the merger has been that, since it now owned only 43.5% of the enlarged group's shares, it no longer had to consolidate the business as anything other than an associate – helpful, given that it reported a loss of INR67,589 million ($975 million) for the year to March 2019. This is an interesting reversal of the strategy employed in the mid-1990s.

CHAPTER 16

Sibling rivalry

O ver the years and in all parts of the world, the mobile market has retained an ability to produce surprises, often on a spectacular scale. The rise and fall of Nokia is one such, as is Vodafone's almost incredible victory in the hostile bid for Mannesmann. What the Chinese have achieved in a decades gives us a third example – more than a billion people connected for the first time in just 25 years.

Recent events in India can now be added to this list – together, the Indian mobile companies operate the world's second-largest national network.

After the launch of service on the subcontinent in 1995, demand for mobile communications began to accelerate. By the end of June 2006 there were more than 100 million mobile connections, and the market was still growing at a phenomenal rate.

The initial decision to award regional licences had resulted in numerous mobile operators springing up, but by 2006 the number had reduced and only five had significant shares of the market. Bharti Airtel, a publicly quoted subsidiary of Bharti Enterprises, enjoyed the number-one position with 23 million customers, just ahead of Reliance Communications with 22.5 million, and the state-owned Bharat Sanchar Nigam Limited (BSNL) with 21 million. Hutchison Essar was fourth with 17.5 million, while IDEA Cellular came in at just under 10.6 million. The remaining 11.7 million connections were unevenly divided between six other operators.

It had taken just over 11 years to connect the country's first 100 million mobile customers; the second 100 million came rather more quickly, in just 15 months. The third 100 million took 11 months, and the fourth, fifth and sixth ever-diminishing time periods. By March 2011 there were over 800 million mobile connections.

Although that number equated to about two thirds of the population, the distribution of SIMs was not at all uniform. A large proportion of the population – much more than a third – remained unserved. The 800-million total included an unquantifiable (but undoubtedly large) number of duplicates, as canny users bought several in an attempt to arbitrage the discrepancies between different operators' voice tariffs.

No fewer than seven operators reported customer bases of more than 50 million and three of them had more than 100 million. Bharti was still out ahead, with 162 million – enough to make it the world's third-largest national operator, after China Mobile and China Unicom.

Throughout the rest of 2011 and into 2012, the industry averaged 10 million new connections a month, raising penetration to over 75% and the market total to 923 million.

Then it started to unravel.

It transpired that there had been some sort of fraud associated with the award of a number of new licences, and several companies that had acquired spectrum in good faith found that they no longer had that spectrum, or indeed any kind of a business. The casualties included some big names – Emirates International, Telenor, Sistema from Russia and the Indian cable-TV operator Videocon. Largely as a result of that, the industry experienced its first-ever reduction in subscriber numbers with the loss of 800,000 connections during the next quarter.

Then, in keeping with best practice, the Indian Department of Telecommunications obliged owners to register their SIMs, in an attempt to identify individual users and thus prevent or at least reduce cybercrime. This culled an astonishing 72.1 million users, taking the total back down to 850 million by the end of 2012.

The attempt to keep a track of SIMs wasn't prosecuted with any real enthusiasm – probably because it was an impossible task – and soon the connection numbers were creeping back up again. By September 2014 the old record of 923 million had been reached once more, and pretty soon, in March 2016, the billion milestone was passed.

Then something else extraordinary happened.

In September 2016 a new company called Reliance Jio entered the market with an offer that was hard to ignore and even harder to match – buy a Jio SIM and get a daily allowance of 4GB of data at 4G speeds, plus unlimited voice and text, entirely free until the end of the year.

The roots of Reliance Jio lay in Reliance Industries, an Indian textiles company founded in the 1960s by Dhirubhai Ambani. After suffering a stroke in 1986, Ambani had handed over control of what was by then a huge conglomerate to his sons Mukesh and Anil. The two, who didn't much care for each other, split the business into two separate companies, Reliance Industries Limited (RIL), controlled by Mukesh; and Reliance-Anil Dhirubhai Ambani (R-Adag), controlled by Anil, which owned one of the largest mobile operators in India, Reliance Communications.

To prevent their continuing mutual animosity from damaging their respective business empires, the brothers agreed that their companies should not compete with each other. A ten-year no-compete agreement was drawn up, but for reasons that are not entirely clear, this only ran for part of its intended life and in late 2010, five years before its scheduled expiry, the arrangement was abandoned. Their two companies issued a joint press release in which they said they were: 'Hopeful and confident that all these steps will create an overall environment of harmony, co-operation and collaboration between the two groups, thereby further enhancing overall shareholder value for shareholders of both groups'. There wasn't much harmony and even less collaboration, but there wasn't outright hostility either – until Mukesh spotted an opportunity which was just too tempting to ignore.

During the course of 2013 the Indian government held one of its regular spectrum auctions, offering parts of the 850MHz, 1.8GHz and 2.3GHz bands for 4G services. An obscure unlisted company called Infotel Broadband Services Ltd was the only company to acquire a full set of licences in all three bands, in all of the 22 'circles' into which the country had been divided. Reliance Industries promptly bought it, renaming it

Reliance Jio Infocomm Ltd.

Ignoring the legacy GSM and W-CDMA technologies that dominated the Indian market, Jio decided to offer voice over internet protocol (VoIP) and data services over a soon-to-be built national 4G network. Distribution wouldn't be a problem – Reliance Industries' retail division was India's largest retailer, with over 3,500 stores nationwide and a presence in some 750 cities.

Under the benign mission 'to connect everyone and everything, everywhere – always at the highest quality and the most affordable price', Mukesh Ambani then launched an unbelievably aggressive marketing campaign. The original free 'Reliance Jio Welcome' offer was followed in early December 2016 with another free offer called 'Jio Happy New Year', effectively extending the period of free service to six months.

Less than a month after the launch, Jio had 16 million users on its network. By the end of 2016, after less than four months, that number had leapt to 72 million. Another offer, the 'Summer Surprise', followed at the end of March 2017, extending the free period for a further three months.

By then Mukesh's new business had overtaken his brother's company, Reliance Communications. Like most Indian mobile operators, it was finding that even the most loyal customers were tempted by the offer of a nearly-free service – what revenue Jio had was coming solely from the sales of SIM cards.

In April, the company connected its 100-millionth customer, after which the regulator obliged it to start charging for its services. Some in the industry thought that this would put a stop to Jio's astonishing growth, but although the rate slowed during July, August and September, the company still added more than 15 million new customers. Jio had achieved the fastest 'free to paid' conversion in the history of the mobile industry.

Average revenue per customer at INR 156 ($2.40) was low by international standards, but above the level reported by every other operator except Bharti. And the network was proving up to the task of handling all traffic it carried – an average of 10GB of data, 626 minutes of voice and 14 hours of video per connection per month. Jio had the lowest call-drop rate, the fastest download speed and, at 100%, the best network availability.

The consequences of Jio's entry into the market are still being felt today. In the three years since the company's launch, there's been a sharp contraction in the number of operators active in the market. Two of the largest, Vodafone and IDEA, have merged; two others, Telenor and Sistema, have pulled out; and two others, Aircel, a company backed by the Malaysian mobile operator Maxis, and Reliance Communications, have gone bankrupt.

From one perspective, the creation of Reliance Jio looks like one of the most spiteful instances of sibling rivalry you're ever likely to see. But looked at another way, it's almost wholly positive. Jio's mission, 'to connect everyone and everything, everywhere – always at the highest quality and the most affordable price', is profoundly benign, and the benefits of the initiative are already evident – most of the population of India now have access to a high-quality communications network at a very low price and they are making full use of it. On average, Jio's customers generate 10.9GB of mobile data each month and 823 minutes of voice traffic. Both numbers compare favourably – more than favourably – with developed-world benchmarks.

Jio's achievement is little short of astonishing, but in some ways, the financials are even more remarkable. Reliance Industries reported its financial results for the year to 31 March 2019, and Jio's contribution to the group is clearly visible in the numbers, having become the third-largest contributor to both turnover (INR136,090 million of a total of INR1,416,340 million) and profits (INR26,650 million out of INR104,270 million). Year on year Jio's revenues have more than doubled, from $3,428 million to $7,099 million, while operating profit has increased by an order of magnitude, from $41.8 million to $451 million. These are astonishing numbers.

The only metrics that might give investors cause for concern are those relating to return on capital employed and return on equity. Jio's assets amounted to INR3,604 billion ($52.0 billion) at the end of March 2019, but the after-tax profit number was just 0.82% of that. Unless profits and profitability can be materially improved – and maybe they can be – the venture will not pay back the amount invested to date until 2140.

CHAPTER 17

Today and tomorrow

Demand for mobile communications has continued to grow throughout the second decade of the 21st century in all parts of the world, to the point that today, as we near the beginning of the third decade, everyone in the developing world who wants a mobile phone has one. Many in the developing world are also connected. Most of us now use our mobile devices to access the internet. We live in a connected world.

But behind the scenes all is not well.

In every developed market, and many others too, the telecoms industry has been subject to some kind of scrutiny for the whole of its existence. Governments set rules that limited the activities of these businesses. Equipment used in the network had to conform to certain rigid technical standards, especially equipment located at customers' premises. Tariffs and charges were controlled.

In the early 1980s privatisation of the telecommunications industry was close to the top of the agenda when Britain's Thatcher government was beginning its campaign to roll back the powers of the state. From very early on it was apparent that it wasn't a good idea to sell off BT, freeing it from the constraints of government, without first putting in place some sort of mechanism to control its behaviour, so the government of the day created OFTEL. It initially aimed to put a reasonably flexible cap on pricing, which would have allowed the newly privatised entity to operate freely and efficiently, and would eventually lead to a genuinely competitive market.

It didn't work out like this in practice, either in the UK or anywhere else. No other regulator shared the vision of a gradual reduction in the level of intervention – they had their powers and they liked using them. So, over the years, instead of less and less regulatory interference, the industry has had to suffer more and more.

The creation in the early 2000s of the European Commission – the executive branch of the European Union (EU),[81] responsible for proposing legislation, implementing decisions, upholding treaties and managing day-to-day business – meant that this unhappy state of affairs was subsequently extended across the whole of Europe, and that has resulted in its being widely emulated elsewhere in the developed world.

In many markets, regulators now control the prices the industry can charge for its services – and this is despite high levels of competition in every major telecoms market in the world.

In the UK and in each of the EU states, regulators have imposed mandatory reductions in termination rates and roaming charges. Termination rates are the amount one company charges another to complete a call that began on that other company's network. Over the last decade or so, most European regulators have required mobile companies to reduce these by 80% or even 90%, to the point where they're close to the charges levelled for terminating a call on a fixed network. This is hardly fair, as the fixed operator needs no special equipment to find its customer – they're at the end of the wire. By contrast, a mobile operator needs to be able to find its customers wherever they might be in the world – and that costs money. These reductions have had clear adverse consequences for the operator: revenue has reduced and profitability has suffered, which in turn has restricted investment in the network and the development of new services.

As for roaming fees, which are additional charges incurred when a mobile customer makes or receives a call in a foreign country, regulators in all parts of the world have required operators to lower the level of these

---

81 The European Union, which came into being in 1993, is a political and economic union of 27 member states located primarily in Europe. It has a standardised system of laws that apply in all member states in those matters where members have agreed to act as one. A monetary union came into force in 2002, composed of 19 EU member states that use the euro.

charges, and today they've been entirely eliminated in the EU. Users are blissfully ignorant of the results of this regulatory interference, which is that if the regulator reduces the amount that can be charged for roaming, operators will look elsewhere to recoup that lost revenue.

Roaming revenues are different from every other type of income the operator receives, because there's almost no cost attached to them. The traveller who arrives at Heathrow and makes a call generates revenue for one of the four UK operators without that operator having had to do anything to attract that revenue – no advertising, no subscriber acquisition costs, no handset subsidy, nothing other than the cost of having a functioning network. Thus it is almost pure profit. This means that if the regulator reduces roaming revenues by say, €10 million, that eliminates €10 million in profit. A typical telco's margin of earnings before interest, tax, depreciation and amortisation (EBITDA) is about 33%, so a €10-million reduction in roaming revenues obliges the operator to raise charges elsewhere by €30 million if it is to offset the reduction.

The EC also controls the companies' abilities to restructure and reshape themselves. It's fair to say that the shape of the European mobile market hasn't changed much over the past two decades. There have been a few deals, but only a handful of these have been especially significant. In part, this is because most operators are either reasonably happy with their current positioning or prepared to live with it and make the best of what they've got.

But there's a second reason: the European Commission has blocked most of the proposals that have been put before it. Some suggest that the Commission's objections have their origins in Three's takeover of Orange in Austria in 2012, which reduced the number of operators in the country from four to three, after which all three raised their prices. The operators claimed that this was as a result not of the diminution of competition, but because prices were too low and returns inadequate thanks to the policy of enforced reductions in roaming rates and call-termination charges – plus the burden placed on the industry by the €2 billion the Austrian Treasury pocketed during yet another spectrum auction in 2013.

It's easy to see what the operators meant. The numbers disclosed by

T-Mobile Austria showed that while subscriber numbers rose from around 3 million to over 4 million between 2006 and 2012, revenues fell from €330 million a quarter to €220 million during the same period, with a predictably adverse impact on profitability.

The EU competition commissioner insisted that Hutchison, the owner of Three, set aside some part of its spectrum to allow a fourth entrant back in the market. It did, but no one was interested in coming into a market where penetration was over 155% and profitability was under severe pressure. As the CEO of Telekom Austria put it, 'Nobody is stupid enough to step into an already ruined market.'[82] Nobody, except a regulator or a commissioner, would expect them to.

The Commission has been somewhat schizophrenic in its approach to consolidation. It allowed Deutsche Telekom to buy 25% of the Hellenic Telecommunications Organization (OTE) in Greece and, with that, assume management control; and it remained unconcerned that by 2018 this had risen to 45%. Telenor was allowed to buy Globul in Bulgaria from OTE in 2013, while in 2015 the Irish arm of Three was allowed to buy O2 Ireland in a move that mirrored the Austrian situation, cutting the number of operators from four to three.

Yet when Three's UK subsidiary attempted to acquire O2 UK in an identical deal just over a year later, it was blocked. The European Commission maintained that assurances from Three's owner, CK Hutchison, 'were not sufficient to prevent' a reduction in innovation and the rate at which network infrastructure was developed.[83] The logic behind the decision was elusive.

Hutchison disagreed with the Commission's ruling and protested in strong terms. It argued that the proposed merger would, in fact, enhance network capacity, as well as coverage and quality. Mobile internet speeds would be improved. As it had also undertaken not to raise prices for five years, consumer interests would be safeguarded. But to no avail.

The quasi-random nature of the European Commission's decision-

---

82 Barker, A & Thomas, D (2014) 'Austrian takeover is a touchstone for telecoms dealmakers', *Financial Times*, 23 February.
83 Statement by EC competition commissioner Margrethe Vestager.

making process was exposed later the same year when the Commission allowed Hutchison to merge its Three Italia subsidiary with VEON's Italian offshoot, Wind. This created a new market leader, which took the name Wind Tre. The enlarged business had 31 million mobile customers, or 32.7% of the Italian market. This wasn't thought to represent an anti-competitive concentration of power.

A comparison with the proposed Three-O2 UK deal is instructive and would also have created a new market leader, but it would have been a monster – a company with a massive 31.15 million customers and a clearly dominant 33.0% share of the UK subscriber base. Obviously, that couldn't be countenanced!

The decisions in Italy and the UK came hot on the heels of another negative verdict. Towards the end of 2014 Telia and Telenor had proposed a merger between their respective Danish subsidiaries. Neither company was especially profitable – in 2014 their EBITDA margins were 11.9% and 12.4% respectively – and neither was especially large. The merged entity would have had a 0.064% share of global telecoms revenues and 0.048% of global mobile subscribers. Still, this, for the European Commission, was too great a concentration of power and the deal was blocked.

Do these people have any idea what they're trying to achieve?

In the last years of the 20th century, the mobile industry was the darling of the financial markets. Now it's largely unloved. Earnings growth slowed, then stopped, then went into reverse, drawn down by regulatory constraints, excessive licence fees, excessive competition and rising capital requirements.

So looked at in one light, it might be thought that the recent history of the mobile industry, from the arrival of the iPhone up until now, has been of a search for something – anything – that would help restore the glamour and excitement of years gone by, the years before the dot.com bubble burst.

A number of operators concluded that disposing of some of its businesses – 'rationalisation of the existing portfolio', in the jargon of the industry – may help improve overall returns. Zain and Orascom spring to mind, but Telia is perhaps the foremost exponent of this approach. In the recent past it has sold more than half of its international operations – its subsidiaries in Spain, Tajikistan and Uzbekistan; three of the four companies held through its Fintur joint venture with Turkcell; most of its shareholding in Turkcell itself; and its entire 44% interest in the huge Russian operator MegaFon. It retains just one Eurasian asset – the diminutive Moldovan company Moldcell – and, that apart, is now entirely focused on Scandinavian and the Baltic states. The capital markets have greeted this interesting initiative with ennui, marking the shares down slightly.

For other companies, the answer has been to keep moving into new or underserved markets where mobile penetration was still at a low level. These operators realised that the mobile market was, in fact, far from saturation, as at the end of 2007 there were still only 3.3 billion users worldwide – less than half the world's population. The developing world still offered operators huge opportunities for growth – mobile penetration there was below 35% on average – if they were careful about choosing where to invest.

A third group of companies decided to double up, integrating their mobile services with fixed and internet offerings, and sometimes TV or some other form of entertainment. This was the approach Vodafone took and is still taking today. But this requires formidable financial resources – fixed networks of any kind don't come cheap and the payback period is inevitably a lot longer than it is in the mobile industry.

And mobile specialists who adopt this approach are likely to meet cable-TV companies coming from the other direction. That industry is facing similar pressures, and some of cable operators are now looking to create similar multi-play options by adding mobile to their portfolios. They have an edge here – bolting on a mobile virtual network operator (MVNO) is a cheap option and can be done in a matter of days.

Some operators have concluded that it isn't enough to offer multi-play services – they also have to own some of the content carried over the network. BT in the UK was one of the first to do this, buying the rights to show Premier League football and some top-drawer English cricket and rugby matches. A BT spokesman explained that it had to move into content of this kind in order to compete effectively against Sky, which had been allowed to create something close to a monopoly in UK sports broadcasting at the same time that BT lost its monopoly on telecommunications.

AT&T has taken this a step further, thanks to its remarkable acquisitions of DirecTV and Time Warner. Facing competition from cable operators, internet service providers, alternative long-distance companies and other mobile operators, the US market leader has struck out on its own towards the goal of complete vertical integration. Today, it's not just the largest telecommunications company in America, it's also the largest pay-TV company and the largest creator of digital content.

All this matters because at the moment the mobile industry isn't really being properly compensated for its efforts – and if it isn't allowed to earn an adequate return on investment, it may stop investing. For example, between 2008 and 2018, global mobile revenues rose from $1.47 trillion to $1.72 trillion, an increase of 15.5%. Operating cash flow (broadly, EBITDA) increased by just 5.7% over that same period, implying a reduction in the margin (from over 37% to under 34%). That may not seem like a catastrophe, but when it's accompanied by an increase in debt from $720 billion to over $1.175 trillion, there's perhaps some cause for concern.

Part of that increase in debt stems from a marked increase in the industry's capital expenditure level, which is rising at more than twice the rate of revenue growth, as operators invest in 4G, 5G licences, 5G equipment and all the rest of it. Annual capital expenditure now equates to almost 20% of annual revenues, which is on the high side. It should also be noted that these reductions in profitability have occurred during a decade when the number of mobile connections in the world grew from 3.5 billion to over 7.5 billion, an increase of nearly 120%.

It's not as if there's a huge concentration of power – only AT&T comes close to a 10% share of total revenues and it's nearly half as large again as the industry's number two, Verizon Communications; and neither is anywhere near as large as Apple. Taken together, the industry's ten largest operators account for only 52% of total industry revenues. No other major industry is quite so fragmented as telecommunications, and few others have to cope with the same level of regulatory interference.

Recognising this, a small number of operators are now looking to pursue an entirely different approach, through diversification into new industries, such as financial services, e-commerce, remote learning, domestic surveillance and the myriad other services that get grouped together under the digital flag of convenience. Some of these areas have greater potential than others. E-commerce is a vast market, for example, but tech companies such as Amazon and Alibaba are already well entrenched and won't be easily shifted.

At the time of writing, Facebook had announced its intention to dive into the murky waters of the cryptocurrency scene with Libra, its own digital currency – and this served to highlight the potential for mobile financial services (MFS), which is, for the moment at least, wide open and there for the taking. And the market is huge. Figures floated about in the press suggest that there are at least 1.7 billion people in the developing world without access to banking of any kind, who are therefore unable to avail themselves of the internet's digital bazaars. Facebook aims to change that and suggests its new Libra service will provide a low-cost, simple mechanism that bypasses traditional banking systems, a mechanism that has already attracted the support of eBay, PayPal, Spotify, Uber and 20 more big hitters.

Some of the industry's current problems are in part self-inflicted. When the first 3G phones began to appear, it was assumed that mobile operators would want to maintain a pricing structure that made it impossible for third parties to offer IP-based voice telephony economically, but in most markets this proved impossible. New entrants into the market, with no legacy voice business to protect, began offering data at prices that made it attractive to use VoIP over mobile networks. This allowed third parties to offer voice services that were, in many instances, cheaper to use than the subscriber's normal mobile voice service – the cost of the data used to make the call was less than the cost of a voice call of the same duration. This was especially true of international calls.

WhatsApp was founded in 2009 by two former Yahoo! employees, Brian Acton and Jan Koum. It began life as a messaging service but after less than a year acquired the ability to transmit photographs, and its user base began to climb rapidly. By early 2013 the company claimed 200 million active users for the app; this had doubled by the end of that year, helped along by the introduction of a voice-messaging service.

In February 2014 Facebook bought the business for $19 billion. At the time it seemed like an astonishing amount of money but as the WhatsApp base continued to grow, the price (a record for a venture-backed start-up) began to look less extreme. By August there were 600 million users; in January 2015 the number hit 700 million; and by April the total was 800 million – double the number at the time of the acquisition just 11 months earlier.

The billionth user signed up in February 2016, while the last number published by the company (in December 2017) quoted a total of 1.5 billion. Critics of the 'high price' paid began to fall silent: as traffic on both the text and voice-message apps grew, it was starting to look as though Facebook had bought the global phone industry for just $19 billion.

WhatsApp isn't alone, of course; there are other such services, including Viber, Telegram and Line, but these – and the others in this space – are rather smaller-scale enterprises. Viber, an Israeli company with links to the military and security forces of that country, is probably the largest of the others, but since it was acquired by the Japanese tech conglomerate Rakuten, it has taken to reporting on the basis of 'unique phone numbers excluding deactivations'. At the end of its last financial year, it claimed 1,095 million such users, but the number is pretty meaningless without any clue as to how many people have disconnected (after having read the small print on the company's terms and conditions of use, no doubt).

Telegram, a Russian company, and Line from Korea claim to have over 200 million and 165 million users, respectively, while Kakao from Japan brings up the rear with 50 million users of its Talk app. (It's interesting, and probably significant, that all of these competitors originate from countries where the written form of the language uses non-Roman characters.)

It's remarkable that a handful of the world's leading internet companies are now valued more highly than the entire telecommunications industry: the so-called BATFAANG group – Facebook, Amazon, Apple, Netflix and Google; plus the three Chinese giants, Alibaba, Baidu and Tencent. But, as noted, these internet businesses couldn't function without the telecoms operators on whose terrestrial networks they piggyback. They rely on the competition between operators and, of course, the helpful intervention of regulators who oblige mobile companies to carry this traffic at near-zero rates. Is this a weakness they can live with? In most cases, it seems the answer is no. Several of the BATFAANG group are actively involved in the construction of an alternative to these terrestrial networks.

The idea of building a constellation of satellites isn't especially new – in 1990 Motorola shocked the mobile industry by announcing its plans to launch Iridium, a new space-based communications network. As we've discussed, the issue with using geostationary satellites for voice traffic is that it takes the signal some time to travel from Earth to the satellite and back again, which tends to make a spontaneous conversation difficult.

Motorola thought up a way around this, proposing that a network be built using a fleet of small satellites, each interconnected with both the terrestrial phone system on Earth and its four immediate neighbours in space, so any call could be handed off in the same way a cellular call was handed off as the edge of a mobile cell was reached.

The design ultimately called for a constellation of 66 low Earth orbit (LEO) satellites some 780 kilometres above the Earth's surface. However, even though it was possible to launch as many as seven satellites from a single rocket, this process of positioning all 66 took from 1997 until 2002.

And the dynamics of the terrestrial mobile industry had changed somewhat by the time the first test call was made in 1998. Terrestrial mobile handsets and the tariffs for their use were a fraction of what they'd been in 1987 when the space-based communications network was conceived. Motorola's proposed call charges were just not attractive – and neither were the handsets, which were bulky, expensive, short on battery life and not particularly effective. The project went bust in 1999 almost immediately after the first call was made over the system.

Other companies also tried and failed: ICO Global Communications, an offshoot of the London-based Inmarsat Group lasted just four years, filing for bankruptcy in 1999; Globalstar, formed in 1991 by the US defence contractor Loral Corporation, followed suit in 2002, while Craig McCaw's

Teledesic never took off, as by the time it was ready to go, the dot.com bubble had burst and funding couldn't be found.

It might be thought that this record of total failure doesn't auger well for the new space-based ventures – but technology has moved on apace, so this new generation of LEOs stands a far better chance of success than their predecessors from the 80s and 90s. The costs of making these satellites have fallen by an order of magnitude (at least) as have launch costs, while the capacity offered by this new generation of LEOs is far, far greater – measured in gigabits, not megabits.

These systems will offer connectivity from a fleet of interconnected vehicles, at least one of which will be 'nearly always nearly overhead' (NANO) at all times, at speeds that rival and, in some circumstances, better those of 'future proof' fibre. There are other advantages too, as Geoff Varrall of RTT explains: 'It is sunnier in space. It does not rain in space. Multifunction solar panels are now achieving 40% efficiency, so that is 20 years of free RF power and no rent to pay. Network densification is also easier (less expensive) in space. (There is more space in space.) It is also cold in space (-270.45°C), so there is no air conditioning to worry about. Radio waves and light travel faster in free space than in a fibre optic cable. Once a fibre optic cable reaches a certain length (about 10,000 kilometres), the free-space advantage outweighs the round-trip distance (1,400 kilometres).'[84]

At the time of writing, the US FCC was considering several proposals to launch constellations of LEO satellites to provide internet connectivity over the entire surface area of the globe.

In May 2019 Elon Musk's SpaceX company launched the first 60 Starlink broadband satellites and plans to add a further 11,840. The mild-mannered South African-born billionaire has also filed an application with the FCC for permission to build as many as a million Earth stations to support the Starlink constellation. One million! Where's he going to put them – as, clearly, they'll need to be everywhere, or at least everywhere where there are potential customers. Is there any organisation anywhere that owns or has access to that much real estate?

---

84 Varrall, G (2018) 5G and Satellite Spectrum, Standards, and Scale. Artech House, London.

The logistical challenge looks considerable, if not insurmountable, until you remember that between them, the four largest independent tower companies owned 340,000 cellular sites at the end of 2018, while China Tower, the largest company of its kind, owns nearly 2 million by itself. It may not be happy to deal with Mr Musk, but someone will be. If all goes well, he plans to have the whole thing up and running by the middle of the 2020s and intends to use the revenues from the venture to fund his plans to build a city on Mars.

Jeff Bezos, the Amazon founder and CEO, has a similar plan for an orbiting broadband network, though his Project Kuiper involves a rather more modest 3,236 satellites spread across three orbital planes, at 590 kilometres, 610 kilometres and 620 kilometres, to 'provide low-latency, high-speed broadband connectivity to unserved and underserved communities around the world'[85] – and, of course, to everybody else, and especially those of us with enough money to buy stuff from Amazon.

But Bezos doesn't just want to operate a system of satellites, he also wants to put them into orbit. Although nothing has yet been confirmed, the Kuiper constellation is likely to be placed in orbit by rockets manufactured by another of Bezos's ventures, Blue Origin, which announced in March 2017 that it would be providing launch services for OneWeb, a third LEO venture backed by Airbus Industries, SoftBank, Virgin and Qualcomm, among others. OneWeb has plans to initially launch and operate a constellation of 650 small satellites although the fleet could eventually run to several thousand.

The Canadian company TeleSat is planning a similar venture, as is LeoSat, a consortium backed by Japan's SKY Perfect and the Spanish satellite operator Hispasat. The Luxembourg-based SES is also involved, through its ownership of O3B, one of the first of ventures to address this opportunity.

Other names that have been mentioned in this context include Facebook, Apple and Boeing – heavyweights all. Google dipped into the industry, acquiring SkySat, a company focused on the business of Earth imaging, for

---

85 'Amazon will launch thousands of satellites to provide internet around the world' on *The Verge*, 4 April 2019.

$500 million in June 2014; although it sold the business to Planet Labs in 2017, it seems improbable that it has no further ambitions in this direction.

What might the telecoms industry look like if Jeff Bezos and Elon Musk get their way? Just as WhatsApp has taken a significant slice out of telecoms companies' traditional traffic revenues, so these new LEOs might have an equally detrimental impact on their income from data and the internet. Bezos already offers certain Amazon customers free delivery of goods; what if he were to extend that altruistic approach to the access to bandwidth on Kuiper? He could recover his costs through increased e-commerce volumes but what could the telecommunications companies do?

There are numerous obstacles for this new industry to overcome – technical, regulatory and financial – so the future will probably be more complicated than that, but such an initiative has the potential to transform everything – not just the communications industry, but the whole way we do commerce.

And what might the telecoms industry look like if Bezos and Musk fail? There's plenty of evidence to suggest that the industry is changing radically, even without extraterrestrial assistance. Are we on the verge of another telecommunications paradigm shift?

On these occasions, people can get caught overextended, doing something that used to work but that has become overly popular. The adverse consequences are invariably considerable: the global financial crisis of 2008 is a case in point. Borrowing heavily to buy property worked for a long time, as property prices appeared to be on a steady and continuous upwards trajectory. People kept gearing their property, in some cases compounding the risk they faced by buying complex derivative instruments. When the paradigm shift eventually occurred, these investors became overextended. There was too much housing supply, not enough housing demand, too much debt, not enough assets, too much interest and not enough yield. The boom came to an abrupt end, lots of people lost lots of money, and we're still living with the effects more than ten years on.

As the American investor and philanthropist Ray Dalio observed in July 2019, 'I have found that the consensus view is typically more heavily influenced by what has happened relatively recently (ie, over the past few years) than it is by what is most likely. It tends to assume that the paradigms that have existed will persist and it fails to anticipate the paradigm shifts, which is why we have such big market and economic shifts. These shifts, more often than not, lead to markets and economies behaving more opposite than similar to how they behaved in the prior paradigm.'[86]

In telecoms there have been many such paradigm shifts. Early electrical technology changed communications from line-of-sight signals to the telegraph. The telegraph was replaced by telephones. Telephones gave way to mobile phones. Mobile phones have now become the mobile internet. As the new 5G technology is being introduced around the world, some network operators seem to assume that today's paradigm will continue to apply, namely, that owning licensed spectrum is the key to wealth creation. Incumbents and new entrants alike are copying the playbook mobile operators have been using since 1993 – getting licensed spectrum, rolling out a network, and sitting back and waiting for the customers and profits.

If only it were that simple. The historic paradigm is shifting from voice to data. The first companies to launch 5G are already beginning to report details of this experience – Korea is leading the charge towards the new standard, and early uptake appears to be positive. However, unlike voice, data is a low-margin product and may not easily support the capital investments that in the past were justified by voice revenues. Less than three months after the launch of service in April 2019, the three Korean operators have connected over 1.3 million customers, equivalent to more than 2% of all subscribers in the country, yet revenues have barely moved. Indeed, the second quarter's numbers are marginally below those seen in the same quarter of the prior year. And profits are down across the board. The new paradigm is low-margin data-only networks.

And what about us? This story is, after all, as much about humanity as technology or industry. At the start of this century, hardly anybody had access to a computer and even those who did were only connected to it for part of the time. Today, almost everyone in the developed world and

---

86 'Paradigm Shifts' by Ray Dalio on LinkedIn, 17 July 2019.

many in the so-called emerging markets carry such a computer with them everywhere they go.

There's no doubt that smartphones have changed us and are continuing to change us. It seems we've become dependent on these little black boxes in half a generation. You can see it all around you today – people are now glued to their phones for most of their waking hours and become nervous if not distraught if distanced from them for any length of time. We walk down streets, bumping into other iZombies as they – and we – stare at the small screen we hold in our hand. In restaurants we sit, barely talking to our companions except to comment on some item our smartthing has presented to us. On our holidays, we lounge around swimming pools with our families, each one of us alone, lost in our own private smartworld.

Through our phones, we're connected to a worldwide web of billions of other people – but also to machines of almost incomprehensible power. Not just some of the time, but all of it. We believe what we read on social networks and even demand 'safe spaces' where we won't encounter these unwanted, alien ideas.

Everything we do is now monitored and remembered, possibly forever. We've surrendered our right to privacy, which we signed away when we agreed to the terms and conditions governing some supposedly free new service. The websites we visit are noted; our thoughts are now predictable and our behaviour monitored. The subjects we Google are recorded. Our calls, texts and emails are tracked and occasionally also recorded. Fitness devices check our biometric data and draw conclusions.

We let others, whose aims and motives we can't know, access our deepest secrets, without question. Why does Spotify need to access the address book on my droid? Why does Google Maps want to look at the photos I've taken or read my emails? Why does Viber want to listen to my phone conversations and access my phone's camera and video recorder? Who gave Alexa and Siri the right to listen to what goes on in my home? What else might they be doing that we don't yet know about?

All of this is done ostensibly on our behalf. And this is just the beginning of a megamorphosis. Futurologists talk of man-machine interfaces and artificial intelligence where humanity is linked to and enhanced by electronics. They point to some imagined time in the next few years where we all play host to cybernautically symbiotic devices – benign, of course – that will allow us to keep pace with the advances in artificial intelligence.

Others argue that we're already servants to these devices, and they seem to be diminishing, not enhancing, our intelligence. Psychologists talk of the need to restrict our screen time and suggest we try to limit our limitless uncontrolled browsing.

What does this all mean and where does it lead? As ever, it could go either way. Those futurologists and forecasters may make their predictions, but the future is by its very nature unknowable. And is it not rash to draw conclusions about humanity's future based on the experience of just a few short years? We now have something wonderful – near-total universal human connectivity – and, used wisely, this is a profoundly positive force.

These are very early days. We don't have to create an Orwellian dystopia just because we're now interconnected. Some difficulty in establishing appropriate ground rules is inevitable, and companies will soon realise that they need to rein back their intrusive behaviour if they are to remain acceptable to their customers.

And the customers themselves will calm down once some of the astonishing novelty of continuous connectivity to information wears off – though it shows no sign of doing so yet. Indeed, our appetite for data seems to be limitless. The numbers used to express this fact are so vast as to be almost incomprehensible: 90% of the world's data has been created in the last two years; the world's approximately 4 billion internet users perform more than 50,000 searches every second. According to some recent estimates, we create 2.5 quintillion bytes of new data every day, which equates to 2,220 Petabytes, or some 22.2 million hours of high-definition video. (Some of this creation is meaningless nonsense; some of it will be offensive to some people; and some of it will have real worth. We should accept that – it's part of the deal.)

And by the time you read this, all these numbers will be out of date, superseded by higher ones. Perhaps much higher. In such an environment, it should be possible for all of today's networks to co-exist. More importantly, it should be possible for us all to enjoy the blessings that flow from freely available information, knowledge and perhaps even wisdom.

From a commercial point of view, it doesn't matter whether the digital future is good or bad. It only matters that it is coming. And that future will run over telecoms pipes and airwaves. Who will win the race to carry that future? The biggest operators, or the bravest?

The major players

Most of the credit for the global mobile industry rests with just a few companies, most especially McCaw, Millicom and Vodafone. This is not to discount the achievements of others, which we highlight over the next few pages.

América Móvil

At the end of 2018 América Móvil was the world's sixth-largest mobile operator, enjoying a dominant position in many of the major mobile markets in Latin America, including Argentina, Brazil, Colombia and Mexico. It also owned a controlling stake in Telekom Austria and a smaller stake in KPN, the former PTTs in Austria and the Netherlands.

América Móvil was formed to bring together the various mobile assets owned by its parent, the national telephone company, Teléfonos de Mexico, or Telmex. Telmex was originally formed in 1947 by a group of private investors to acquire the telephone network in Mexico owned and operated by LM Ericsson, the Swedish manufacturer of telecommunications equipment, which at the time had interests in a number of overseas networks. In 1950 Telmex bought the other telephone network in the country, a subsidiary of American conglomerate ITT. It operated as a privately owned de facto monopoly until 1972, when it was nationalised by the Mexican state.

The telephone service was truly appalling in the years following the nationalisation: making a call might involve travelling to the local exchange building and a long wait for one of the very few lines installed there to become available. Whether this would lead to an actual connection was always in doubt, and even if it did, there was no guarantee that the signal would be audible. This clearly had to be rectified, and in 1990 the Mexican government decided to sell a controlling stake to a consortium of investors led by Carlos Slim's Grupo Carso and including France Telecom and SBC, the Bell company based just over the border in neighbouring Texas.

The introduction of new management and new capital led to dramatic improvements in the quality of the service and the size of the network, which in turn facilitated a period of considerable economic growth. Telephone penetration rose from 6% in 1990 to over 9% by 1994. At the same time, the proportion of lines served by digital exchanges more than doubled, to over 80%, and revenue from telephone services also more than doubled. Mobile service was initiated in 1989, fuelling progress at a similar rate for the rest of the decade and beyond.

As the 1990s progressed, both BellSouth and Telefónica, the Spanish telephone company, were acquiring equity stakes in mobile businesses in Latin America. Success of this kind invariably attracts attention, and Telmex decided it too would like to have some involvement in these markets. Under normal circumstances, a third-world PTT wouldn't find backing against this kind of opposition, but Telmex wasn't just any old monopolist – Carlos Slim was a man of vision, energy and determination.

By the late 1990s, the telephone market in Brazil was undergoing reorganisation and restructuring. Slim recognised that this represented a huge opportunity for Telmex to expand the scope of its operations, and looked for ways to participate in the process and acquire an interest in some of the new operators. On its own, Telmex might not look an especially attractive partner, so Slim looked to the north, to one of his original partners in the privatisation process, SBC. Together, they formed a consortium called Telecom Américas, bringing in two Canadian partners, Bell Canada International and the small independent TIW. Telecom Américas acquired stakes in two operators in late 1999, Algar Telecom Leste and Telet, which operated in Rio de Janeiro, Espirito Santo and Rio Grande do Sul.

However, Brazil wasn't the only target for Telmex. During 1999 it took control of Telecomunicaciones de Guatemala, acquiring an 81.3%

stake during a controversial privatisation; and minority stakes in SBC's Puerto Rico Cellular subsidiary and two of the new mobile companies in Colombia. Slim create a new firm called América Móvil, which would be spun off as a separate listed company, to own these mobile interests.

América Móvil started life with the Mexican mobile business, plus the investments in Brazil, Colombia, Guatemala and Puerto Rico, and an interest in the US mobile market – in 1999 Telmex had acquired the Florida-based Topp Telecom, a company specialising in reselling mobile service to Spanish speakers in the US. América Móvil renamed this TracFone, and today it's the largest reseller of cellular service in the USA.

By 2001 it was clear that América Móvil intended to challenge Telefónica for market leadership in Brazil. It bought controlling stakes in several of the other non-wireline mobile operators during the first years of the new century, bringing it within reach of full national coverage. By the end of 2002 it had acquired additional licences, covering the Sao Paolo state, Santa Clara and three other regions. It had bought out its original partners in the Telecom Américas consortium.

The year 2003 was characterised by frantic activity. A GSM network had been launched in Mexico in October 2002; the same technology was introduced to Ecuador and Colombia in 2003. The first acquisition of the year took place in February 2003. Celcaribe, a business covering the east coast of Colombia, was bought from Millicom, giving América Móvil a complete national presence. March saw another piece of the Brazilian jigsaw fall into place when América Móvil bought BSE, a company licensed to operate in the north of Brazil owned by BellSouth and its local partner, Grupo Safra. Later in 2003 América Móvil took control of another BellSouth investment, BCP, an operator with a licence that covered metropolitan Sao Paolo.

These acquisitions marked a turning point in the development of the telecoms industry in Latin America – BellSouth was moving out of the continent and América Móvil was taking over. Five other deals were completed in 2003, taking the total for the year in Latin America to eight. These strengthened the company's presence in Ecuador and took it into another three new markets, Argentina, El Salvador and Nicaragua.

The pace slackened slightly in 2004 and 2005, but there were still deals to be done. América Móvil bought Megatel, a business in Honduras, in June 2004, and acquired additional shares in Compañía de Telecomunicaciones de El Salvador (CTE), the company it had bought from France Telecom the previous year. In 2005 it bought Hutchison's business in Paraguay, following this with acquisitions in Chile (Smartcom) and Peru (TIM Peru). The following year it bought stakes in three businesses from Verizon – mobile business in the Dominican Republic and Puerto Rico and a minority stake in Compañía Anónima Nacional de Teléfonos de Venezuela (CANTV).

The footprint América Móvil had begun to create in Brazil was finally completed in September 2007, when the company was awarded licences to operate in Minas Gerais and the northern states. It remains one of just four operators with coverage of the whole of Brazil.

It seemed as though most of the hard work had been done – but by this time, América Móvil was planning something rather more significant. In January 2010 it bid for control of Carso Global Telecom, the company that the Slim family had created to hold its interests in Telmex. By June, Telmex and Telmex International had become subsidiaries of América Móvil. The continental footprint was now close to completion.

The award of new licences in Panama and Costa Rica and the acquisition of Digicel's subsidiaries in Honduras and El Salvador brought an end to this phase of the company's expansion. Not even Vodafone had managed to establish such a concentrated presence: by any standards it was a remarkable achievement.

In May 2012 América Móvil surprised almost everybody by launching a partial tender offer for KPN, the Dutch national phone company. The financial markets took a dim view of the deal; América Móvil's own stock dropped by nearly 10% the following day and continued downwards for another eight trading days. Undeterred, Slim ploughed on and by the end of June the primary objective had been achieved: América Móvil was now KPN's largest shareholder, having acquired a 27.7% stake at a cost of about €3 billion.

The markets reconsidered the deal – and still didn't like it. KPN had, it seemed, tried to improve its chances of remaining independent by offering to sell its best international asset, E-Plus in Germany, to Telefónica, but that had failed because the Spaniards were also in a difficult financial position. In the end, KPN had to rely on a 'poison pill' to fend off América Móvil and maintain its independence. The KPN Foundation, which was established to safeguard shareholders' interests, announced that it would be exercising an option to acquire new preference shares. This effectively blocked the road to control.

While all this was happening, on 15 June, América Móvil announced that it had agreed to acquire a 21% stake in Telekom Austria. That announcement was better received – the former monopoly hadn't been burnt in the 3G auctions a decade earlier and had built up a sizeable presence in the mobile markets of Eastern Europe, with wholly owned subsidiaries in Belarus, Bulgaria, Croatia, Macedonia, Serbia and Slovenia, as well as a small operation in Liechtenstein. The state still owned a 28.4% interest in the business, but didn't seem to have any qualms about letting the Mexicans take control. By the end of September, América Móvil had raised its stake to 22.76% through further purchases in the open market. Two years later, it tendered for the rest of the shares not held by the state and won control, with 50.8% of the equity.

Over the last five years América Móvil has continued to acquire businesses as and when they become available, but the deals are now tactical rather than strategic. In early 2019, it bought Nextel International's small Brazilian business and two Telefónica subsidiaries in Guatemala and El Salvador, but neither changed the shape of the business much.

AT&T

The world's first telephone company was founded by Alexander Graham Bell, the Scotsman who invented the telephone. It was incorporated as the Bell Telephone Company in July 1877.

Under the inspired guidance of Theodore Vail, the new business grew rapidly, awarding licences to independent companies to operate local networks in parts of the country it didn't serve. Crucially, Bell retained its monopoly on all long-distance traffic – the part of the market that was the least capital intensive and most lucrative.

It moved into manufacturing with the acquisition of Western Electric in 1881, after which it took on a new name, American Telephone and Telegraph Company (AT&T). International subsidiaries were created, and by the end of the 1880s the telephone had become established in Europe, Japan, South Africa and elsewhere.

By 1914 nearly 10% of all American homes had access to a telephone.

In 1918 the US government chose to nationalise the company, making it a branch of the US Post Office. The following year, this decision was reversed after widespread public protest.

The 1929 Wall Street crash brought a stop to the seemingly inexorable growth in telephone use, but the downturn was only temporary. Growth resumed in the late 1930s and began to boom again during and after the Second World War. This was a time of great innovation. In 1946, AT&T began offering the world's first car-phone service, and the following year it defined the cellular-radio concept and also demonstrated the world's first working transistor.

Throughout the 1950s and early 60s pressure was building on the company to relax aspects of its monopoly. In 1963, a small start-up called Microwave Communications Inc. (MCI) was formed to bid for a number of licences to operate microwave relays in competition with AT&T. In 1969 the FCC awarded MCI a number of licences that would allow it to offer long-distance service over those links in direct competition with AT&T.

MCI wanted more. In 1974 it sued AT&T under antitrust legislation and won. The court case led to AT&T's decision in the mid-1980s to break up, separating the local exchange business from AT&T's manufacturing and long-distance businesses. AT&T spun off its local exchange business to seven new regional companies – the 'Baby Bells'. Crucially, AT&T decided that it wasn't interested in the new cellular licences that were being awarded to the local telephone operators, preferring instead to retain the lucrative Yellow Pages directory business.

In 1994, it realised that the decision to ignore mobile had been a bad one and, at a cost of $11.5 billion, bought McCaw Cellular, the largest independent mobile company in the US.

As the end of the millennium neared, AT&T seemed to be in two minds about what direction to take. It had spun off its manufacturing side, Lucent Technologies, in 1996, and also the computer interests it had acquired through NCR Corporation, the old National Cash Registers. In March 1999 it moved into the cable business, buying Tele-Communications Inc. (TCI), the largest cable-TV operator in the US. In the same year it set up a tracking stock for AT&T Wireless, which created a psychological distance between parent and subsidiary, even if there was no actual separation at that time.

In 2001 AT&T disposed of AT&T Wireless and Liberty Media, a content company that had been a subsidiary of TCI. Later that year it effectively disposed of the rest of TCI when it merged the business (by then renamed AT&T Broadband) with Comcast.

In November 2005 SBC Communications acquired what was left of its former parent for $18 billion.

AT&T (formerly SBC Communications)

At the time the original Bell System was dismembered, the company now known as AT&T had the name Southwestern Bell (SBC). Today, this giant business owns three of the other regional Bell companies (Ameritech, Pacific Telesis and BellSouth); its one-time parent, the AT&T long-distance business; a number of smaller US telephone companies; DirecTV, the world's largest pay-TV operator; and Time Warner, the world's largest media company.

SBC Communications

SBC was based in San Antonio, Texas, when it was spun off from its former parent in January 1984. It operated a wireline business in five American states (Arkansas, Kansas, Missouri, Oklahoma and Texas) and held the B-block wireline licence for numerous large markets in those states. It was one of the first companies to realise that it could make cellular acquisitions outside its traditional market area, and by the beginning of 1994 it had franchises in 28 MSAs, including five of the top 15 – Washington, Chicago,

Boston, St Louis and Dallas.

In 1997 SBC acquired Pacific Telesis (PacTel), the regional Bell company that operated in California and Nevada. PacTel was largely a wireline business by that time, as it had divested its cellular interests in 1993, spinning these off as AirTouch.

In 1999 SBC repeated the trick, buying Ameritech, another of its sister companies. This acquisition took SBC into the wireline business in five midwestern states and several Midwest mobile markets, including Detroit and Milwaukee.

As the market went digital, SBC adopted the GSM standard, and in 2000 it completed a joint venture with BellSouth, called Cingular Wireless. This filled many of the remaining gaps in its near-national footprint. By the end of that year it was providing service in eight of the top ten MSAs and 33 of the top 50.

Cingular acquired AT&T Wireless in 2004 to create a new leader in the US mobile market. SBC acquired its former parent AT&T in 2005 and changed its own name to AT&T. In 2006 it bought BellSouth. The original monopoly was slowly reassembling itself – four of the seven original regional Bell companies and the AT&T long-distance business had now been brought back under common ownership.

When AT&T attempted to buy T-Mobile in 2011, the US Department of Justice blocked the deal but AT&T was soon back on the aquisition trail. In July 2015 AT&T acquired DirecTV and at a stroke became the largest pay-TV business in the world. The following year it took control of Time Warner and became the largest media company in the world.

SBC's international expansion plans were initially somewhat tentative – it began by acquiring minority stakes in businesses rather than taking control. In 1990 it took a stake in the privatisation consortium in Mexico through which it acquired a 10% interest in Teléfonos de Mexico, the former PTT, which was in dire need of investment and modernisation. Teléfonos de Mexico generated fabulous returns over the whole period during which SBC was an investor, as it expanded across the whole of Latin America.

Next, SBC became AirTouch's partner in South Korea, where the two won stakes in Shinsegi, the shortlived second operator.

In 1994 SBC paid $626 million for a 22.22% interest in Transtel, a holding company that CGE had created to effect a reorganisation of its French telecom interests. The new entity's principal asset was a 50% stake in Cegetel, which in turn owned 59% of COFIRA, the vehicle created to own Société Francaise de Radiotéléphones (SFR), France's second mobile operator. The 22.22% interest equated to an indirect stake in SFR of precisely 10%. During this reorganisation, BellSouth sold its 4% stake in COFIRA, Vodafone swapped a similar 4% stake for a 10% direct interest in SFR, and BT and Mannesmann came in as 26% and 15% shareholders in Cegetel. Several other smaller investors also sold out.

The next deal was rather more straightforward. In early 1995 SBC acquired a 40% interest in Vía Trans Radio Chilena Radiotelegraphy Company (VTR) Celular, an independent mobile operator in Chile, for $317 million. (VTR Celular had begun as a joint venture between VTR Telecommunications and Millicom in 1991, licensed to provide mobile services across the country except in Lima and Valparaiso, the same target market VTR had originally been set up to serve back in 1928. Millicom sold out at the beginning of 1994, so a new partner with mobile expertise was required.)

Later in the year, SBC made an inspired investment, acquiring a 15.5% interest in M-Cell in South Africa, the company that would eventually become MTN. This had been formed by Cable & Wireless in partnership with various local investors, but right from the start it had been a poor second to Vodacom, the country's other GSM operator. SBC bought in at a price so low that it didn't bother reporting it to its shareholders – just $90 million.

MTN has gone on to become one of the industry's greatest successes, but unfortunately SBC chose not to go along for the ride: in 1997 another, superficially more attractive, opportunity arose in South Africa that created a conflict of interest. SBC formed the Thintana consortium with Telekom Malaysia (60:40) and at a cost of ZAR5.16 billion ($1.24 billion) took a 30% stake in Telkom, the South African telephone monopoly. Because Telkom was the largest shareholder in Vodacom, M-Cell's mobile competitor, the deal obliged SBC to sell its interest in MTN. Thintana made money for

SBC, but it would have made much more had it retained the MTN stake instead.

International assets were never really more than an interesting distraction for SBC. After the May 1998 merger with Ameritech, the various overseas investments were considered inessential and, one by one, SBC sold them to focus on the US.

Ameritech

American Information Technologies Corp (Ameritech) was the name taken by the regional Bell company operating in the Midwest. Ameritech's service area covered five states: Illinois, Indiana, Michigan, Ohio and Wisconsin. In 1983 it became the first of the seven regional Bell companies to launch a cellular service.

It wasn't especially interested in expanding its US presence, preferring to concentrate on its traditional service area, but it did acquire a number of international investments. The first of these came in 1990, when it formed a joint venture with Bell Atlantic to acquire 49.75% of the Telecom Corporation of New Zealand (TCNZ), the New Zealand PTT.

Another partnership was formed the following year, this time with France Telecom (as it was at the time) to invest in CenterTel, a mobile network owned by Telekomunikacja Polska SA (TPSA), the Polish telephone company.

Over the next couple of years Ameritech bought into NetCom, Denmark's second GSM operator; Matav, the Hungarian PTT (in conjunction with Deutsche Telekom); and Belgacom, in partnership with Singapore Telecom and TDC of Denmark.

In 1998 Ameritech agreed to a merger with a subsidiary of SBC and after a period of regulatory scrutiny, became a subsidiary of its sister company.

BellSouth

BellSouth was the largest of the seven regional Bell companies at the time of divestment, operating wireline systems in Alabama, Florida, Georgia,

Kentucky, Louisiana, Mississippi, North Carolina, South Carolina and Tennessee.

It was the first of the regionals to realise it could acquire cellular companies both inside and outside these nine states: as early as 1984, it established new cellular joint ventures with other local-exchange operators in Chattanooga and Memphis, Tennessee. In 1986 it bought a 15% stake in the Mobile Communications Corporation of America, a company that had investments in some key cellular markets, including Los Angeles. In 1988 it bought the remaining 85% of the business.

By this time it was beginning to look overseas. In 1988 it joined the COFIRA consortium that CGE was assembling in France, and acquired a 35.9% interest in Compañía de Radiocomunicaciones Móviles (CRM), an Argentinian mobile company licensed to provide mobile services in Buenos Aires.

In 1989 BellSouth was awarded a mobile licence in New Zealand, the country's second, but the first to use GSM technology. In the same year it bought 35% of Abiatar, a mobile operator in Uruguay, and took a 36.4% stake in a consortium that won a regional cellular franchise in the Mexican state of Guadalajara. In 1991 it strengthened its position in Latin America by acquiring a 44% stake in TelCel in Venezuela, then bought 100% of Cidcom, a cellular operator in Chile. Finally, it took a 29% stake in Dansk Mobiltelefon, a Danish consortium that also included AirTouch.

After this, BellSouth was invited to join a consortium that Cable & Wireless was assembling in Australia to bid for Aussat, the state-owned satellite communications carrier. Aussat's main attraction was that it had been granted a licence to operate a GSM system in competition with Telecom Australia (later Telstra), the national PTT. The consortium was successful and in 1991 it began operating with a new name, Optus Communications.

In 1993 BellSouth joined forces with British company Vodafone to win the first GSM-1800 licence in Germany, and also joined the SkyCell consortium that successfully acquired a GSM licence in the Indian city of Madras (now Chennai).

In 1994 it sold its French and Mexican assets.

In 1995 it acquired a 30% interest in Cellcom, a consortium that was awarded the second licence in Israel.

In 1996 it successfully bid for a build-operate-transfer (BOT) licence in Panama.

In 1997 the company acquired a 59% stake in Tele2000, a mobile operator in Peru that had a licence to operate in Lima and Callao, one of the larger provincial cities. Then it moved into Brazil, taking a 41% interest in a consortium (BCP) that was to win the second mobile licence in Sao Paolo, and a 42.5% interest in a similar company (BSE) that won a licence in the northeast region.

In 1998 BellSouth invested in a fixed wireless business in Ecuador and a mobile operator in Nicaragua.

Towards the end of 1998 BellSouth sold its 65% stake in New Zealand to Vodafone, the first real indication that its focus was shifting back towards the Americas.

In 2000 BellSouth agreed to merge its domestic cellular interests with those of SBC to form Cingular Wireless. When, in 2004, Cingular went on to bid for AT&T Wireless, BellSouth needed to raise cash to fund its share of the deal. It had already sold its interests in Brazil to América Móvil in 2003, so the only real option was to sell the whole of the rest of its extensive Latin American portfolio to América Móvil's fiercest rival, Telefónica.

In December 2006 AT&T, the former SBC, acquired BellSouth, bringing the two neighbours together again after nearly a quarter of a century apart.

Cingular Wireless

Cingular Wireless was created in October 2000, when SBC and BellSouth agreed to merge their US mobile business. SBC became the senior partner, with 60% of the new business.

In February 2004 Cingular acquired AT&T Wireless to create the largest mobile operator in the US. Following this deal, Cingular also adopted the AT&T brand.

AT&T Wireless (formerly McCaw Cellular)

In 1994 AT&T acquired McCaw Cellular Communications for $11.5 billion. The company was ten years old when AT&T acquired it and had accumulated an enviable portfolio of wireless assets in the US covering almost all of the main markets. It had been the first company to grasp the importance of the new technology and took very considerable risks to build up its presence in the market, doing this with the use of high-yield debt – the notorious junk bonds pioneered by the even more notorious Michael Milken. For AT&T, it was worth every penny, as it reversed the stupid decision not to bother with the mobile market that it had made before the 1984 break-up.

The McCaw footprint was strengthened through aggressive participation in the 1994/95 PCS auctions, during which AT&T acquired licences covering a further 77 million pops for a total of $1.68 billion. AT&T began bundling cellular with its existing long-distance service, offering discounts on long-distance rates and handsets for just $1. However, mismanagement caused an exodus of senior staff from company, leaving AT&T staffers in charge of both the day-to-day management of the business and its strategy. The cellular company began to wither on the vine.

An attempt to revive it was made in 2000, when AT&T sold 15.6% of the equity to public investors in what at the time was the world's largest-ever IPO. In January 2001 NTT DoCoMo acquired a 16% stake in the company from AT&T, reducing its holding to 68.4%. In July that year, AT&T got out of the mobile business for the second time in less than 20 years, spinning off AT&T Wireless to its shareholders.

In 2003 the US FCC eventually got around to introducing number portability in the US, giving customers the right to take their cellular number with them if they changed network. While the move didn't produce the mass exodus as some had anticipated, it did require AT&T Wireless to spend heavily on customer acquisition and retention, and this pushed the business back into loss.

In early 2004 the one-time market leader gave up the struggle and put itself up for sale. It was bought by Cingular Wireless.

Pacific Telesis

Pacific Telesis (PacTel) was one of the smaller regional Bell companies with a footprint that covered just two states, California and Nevada, and even there it didn't enjoy a totally dominant position – in 1982 AT&T had allowed GTE to control the wireline cellular licences in San Francisco and San Jose, two key markets. The annoyance felt by PacTel executives about this curious decision helps explain the considerable effort the company made to diversify – both regionally within America and internationally.

In 1986 PacTel acquired a stake in Communications Industries, the non-wireline operator in both those cities (and elsewhere), after which there was no stopping them.

Bolstered by the strength of the local phone business, PacTel bought numerous domestic cellular businesses, which took it into such key markets as Detroit, Cleveland, Atlanta and Dallas/Fort Worth. Under the inspired leadership of Jan Neels, PacTel International acquired an outstanding portfolio of mobile businesses in Belgium, Denmark, Germany, Italy, Japan, Portugal, South Korea, Spain and Sweden.

In 1993 PacTel concluded that it could create greater value for its shareholders if these assets were separated from the over-regulated local telephone business, and it subsequently spun off its entire cellular business, domestic and international, into a new company called AirTouch Communications.

The PCS auctions in 1994/95 provided PacTel with a way back into the mobile business in those markets that mattered most, and the company acquired spectrum in Los Angeles, San Francisco, San Diego and San Jose, cities with a combined population of over 30 million. Shortly after this, in April 1996, it accepted a $16,5 billion takeover offer from SBC, its sister company.

Verizon Communications

Verizon Communications was formed by the merger of Bell Atlantic and GTE in 1999. Its principal mobile subsidiary, Verizon Wireless, was the result of a three-way merger between Bell Atlantic's mobile subsidiaries, those of GTE and the US mobile assets owned by Vodafone AirTouch.

In the years since that deal, Verizon has focused on its huge domestic market and has generally preferred to grow organically, rather than through acquisition. That said, there have been two notable deals over the last 20 years. The first, in 2005, was the acquisition of MCI, the long-distance specialist that had been the catalyst for the dismemberment of the Bell System back in 1984. The second was the acquisition of Vodafone's 45% interest in Verizon Wireless.

Bell Atlantic

Bell Atlantic was one of the seven regional companies created by the breakup of AT&T in 1984. It acquired the local Bell telephone companies in the states of Delaware, New Jersey, Maryland, Pennsylvania, Virginia and West Virginia, and also in the capital, Washington DC.

Bell Atlantic Mobile announced the acquisition of Metro Mobile CTS in 1991. It followed this in 1993 by bidding for TCI, the largest cable company in the US. The move surprised, even shocked, investors at the time – the proposed $21-billion deal would have created a giant corporation capable of offering triple or even quad-play services. Just a few months later, however, regulatory intervention from the FCC undermined the economics of the deal and the two called it off. Six years later, TCI would be bought by AT&T, Bell Atlantic's former parent.

At the time of the 1984 break-up, a number of commentators suggested that it was a mistake to split the eastern seaboard into two. The area from Massachusetts down to Washington (and perhaps beyond) was really just one large market. But a company that operated across the whole of that territory would be huge – far too large to meet the criterion of being smaller than the largest of the independents, GTE. Thus, Bell Atlantic and NYNEX to its north had been set up as separate entities. In 1995 Bell Atlantic and NYNEX took a first step towards rectifying this when they merged their mobile businesses; two years later they merged completely under the name Bell Atlantic.

Long before this, Bell Atlantic had made its first move outside the US – in 1990 it had established a joint venture with Ameritech to acquire a 49.75% stake in TCNZ. In the same year it had allied itself with another of its sister companies, US West, creating the Atlantic West consortium in order to acquire stakes in mobile businesses in the Czech Republic, Hungary and Slovakia.

In 1993 Bell Atlantic had become a founder member of Olivetti's Omnitel consortium in Italy, with a 16.6% stake, which was diluted to 11.6% after the last-minute merger with Pronto Italia. And finally, also in 1993, it had bought into Iusacell in Mexico, acquiring an initial 23% interest, before raising this to 42%. All of these assets except Omnitel were to be deemed 'non-core' after the 1999 merger with GTE, and one by one the newly formed Verizon disposed of them.

GTE

GTE was formed in 1918 to provide local telephone services in southern Wisconsin. Over the decades, it grew steadily and by 1955, it had become the largest independent operator. GTE was involved in international markets long before the arrival of mobile phones, with fixed-network franchises in the Dominican Republic and two Canadian regions, British Colombia and Quebec. In 1988 it became one of the first US operators to invest in a foreign privatisation, when it headed the Venworld consortium that took a 40% stake in CANTV, Venezuela's national phone company. Its partners included Telefónica from Spain and, with a 5% interest, AT&T.

In 1991 GTE acquired another large independent operator in the USA, Continental Telephone (Contel). This company had a substantial US mobile business, but had also begun investing overseas, teaming up with McCaw Cellular to buy stakes in one of the nine regional Mexican licences that covered the states of Sinaloa and Sonora.

Later that year GTE became an early investor in the Tu-Ka group of companies in Japan. Like most of the international operators involved in these businesses, it was offered only a small fragment of the equity – 3%–4.5%.

In early 1994 GTE headed the Compañia de Teléfonos del Interior (CTI) consortium (which also contained AT&T), which was successful in its bids for licences in the northern and southern regions of Argentina. These allowed it to operate as a monopoly anywhere in the country outside the

Buenos Aires region for two years.

GTE's final international deal was struck in 1997, when it took a stake in the Pacific Communications consortium in Taiwan, one of three companies to be awarded a national mobile licence. This company is now called Taiwan Cellular.

The merger of Bell Atlantic and GTE in 1999 shifted the emphasis back to the USA, and most of these international assets were divested in the early years of the 21st century.

NYNEX Communications

The New York and New England Exchange Company (NYNEX) was formed to acquire the Bell System's assets in New York and five of the six states that constitute New England – Maine, Massachusetts, New Hampshire, Rhode Island and Vermont. (The sixth, Connecticut, was served by Southern New England Bell, which despite its name wasn't part of the Bell System.)

Although NYNEX didn't make any significant acquisitions of domestic mobile assets, it did make a couple of forays into the international sphere. In 1990 it joined Daimler Benz to bid for the D2 licence in Germany, but without success. In 1993 it had better luck, taking a 20% stake in STET Hellas, the winner of one of the two GSM licences awarded in Greece. It acquired and then sold a 23% stake in Indonesian mobile business Excelcomindo (now XL Axiata). Finally, it had shares in several of the Indian cellular regions, and fragments of the equity in the Tu-Ka and Digital Tu-Ka consortia in Japan.

NYNEX and Bell Atlantic merged their domestic cellular businesses in 1995 and merged the rest of their respective companies in 1997.

US West

US West inherited by far the largest share of the US landmass when the Bell System was broken up. Its franchise area covered about 45% of the area of the country (excluding Alaska and Hawaii) but most of the 14 states it served were sparsely populated and essentially rural. There are only a few large metropolitan areas in the 14 – Phoenix in Arizona, Denver in Colorado, Minneapolis/St Paul in Minnesota, Portland in Oregon, Salt Lake City in Utah and Seattle in Washington were the only places with populations of a million or more. Most of the cities in the other states – Idaho, Iowa, Montana, New Mexico, Nebraska, North Dakota, South Dakota and Wyoming – were barely more than large towns.

The company decided it had to be different – and to move fast. It realised that its traditional 14-state telephone business was unlikely to produce much in the way of profit growth, so it began to place the emphasis on new markets – domestic mobile and entertainment services, real estate, and international markets. Between 1989 and 1996 US West acquired mobile interests in Europe in the UK, the Czech Republic, France, Hungary, Poland and Slovakia, plus a 20% stake in Malaysia's Binariang (now Maxis) and several stakes in regional mobile companies in India, Japan and Russia.

US West acquired numerous mobile licences in Russia during the first years after the collapse of the Soviet Union in the early 1990s. The potential of these assets was well understood, but so too were the risks. There was no guarantee that the 800MHz or 900MHz band would be available – in many cases, it was still being used by the remains of the Soviet military – so US West took the decision to put most of the better assets into a separate vehicle, the Russian Telecom Development Corporation. The idea was that this would be run like a closed-end fund, with any profits reinvested into similar Russian ventures until the business had created some real value, after which its future would be reassessed. That point was never reached.

At the same time, US West actively pursued opportunities in cable TV. It was an early investor in the UK, creating the TeleWest joint venture with TCI, its near neighbours in Colorado. TeleWest had accumulated the best collection of UK cable franchises (there were, initially at least, 132 of these) and was licensed to offer both television and telephone services over what at the time were state-of-the-art hybrid fibre-coax networks (a combination of fibre-optic and co-axial copper cabling). This experience proved invaluable and US West later brought it to bear in its own domestic market.

In 1994 US West merged its domestic cellular businesses with those owned by AirTouch to create a business with a powerful presence across the whole of the American west. The following year US West was one of four partners (with AirTouch, Bell Atlantic and NYNEX) in the Prime PCS consortium, designed to acquire licences in markets where none of the four had an existing presence, thereby creating a full national footprint that all the partners could access.

US West was as interested in cable as it was in mobile, and in 1996, having already acquired a 25.5% interest in Time Warner Entertainment, it merged with Continental Cablevision, the second-largest US CATV business.

The foreign cellular assets and US West's numerous cable and entertainment assets were then grouped together in a new holding company, MediaOne, which was initially a wholly owned subsidiary (with a tracking stock) but was spun off completely in July 1998. The following year, AT&T acquired MediaOne's US cable assets, while Deutsche Telekom bought most of the remaining international businesses. The domestic wireline business – the original Baby Bell company – was acquired by Qwest in June 2000, while the remnants of the impressive set of assets US West had accumulated in its relatively short life ended up scattered around the rest of the industry.

Bharti Airtel

By the end of the 1980s the Indian government had realised that its economic policies had left it far behind the rest of the Asia Pacific region, and it determined to do what it could to rectify this as rapidly as possible. Free-market policies were introduced as a part of a drive towards the creation of an industrial society, foreign-ownership rules were eased, foreign investment was given a guarded welcome, and attempts were made to foster the growth of an indigenous technology industry.

Obviously, telecommunications would have to play a central part in this but the country's phone system was in a dire state. At the end of 1990 there were just over 5 million exchange lines serving a country with a population of 845 million. Five years later the number had more than doubled to just under 12 million, but by then the population had risen to 930 million, so telephone penetration was still pitifully low at just under 1.3%. Phone access was also very inequitably distributed: more than a third of all lines were in the main metropolitan areas, where just 8% of the population lived, leaving the rural population with just one line for every 200 inhabitants.

The government had resisted attempts to introduce foreign capital into the fixed network, as this was deemed to be a key strategic asset, but was more amenable to the idea of foreign investment in mobile communications. In 1994, the Indian Department of Telecommunications set a target of 10% penetration by the end of the decade – a further 88 million phones, or the same number of lines AT&T had had in 1983 just before it was split up. That had taken more than a century to create, so India's five-year plan was probably over-ambitious.

The Department of Communications settled on a combination of mobile and fixed wireless, opting to use the GSM standard for the first and CDMA for the second. It invited private investors to bid for two licences to operate new mobile systems in each of four large urban areas, Bombay (Mumbai), Calcutta (Kolkata), Delhi and Madras (Chennai). Local companies were required to own a minimum of 51% of each bidder (though this so-called 'local' company might be a joint venture in which a foreign investor had a 49% interest, allowing the foreign investor a maximum 73.99% economic interest).

The licensing process wasn't entirely free from controversy but when the dust had settled, the list of winners read like a who's who of the world telecom industry – AirTouch, Bell Atlantic, BellSouth, France Telecom, Hutchison, Millicom, Mobile Systems International (MSI), Telekom Malaysia, Swisscom, Telstra from Australia, Vivendi, and CCI International, a subsidiary of Cellular Communications, AirTouch's partner in certain US franchises.

The country's first mobile network was launched in Chennai in 1995, by RPG Cellular, a company controlled by an Indian conglomerate but in which Vodafone had a 20% interest. Over the next few months all the others entered the market and by the end of that year there were 30,328 customers connected. One year on, the number had risen to over 235,000 – still only 0.02% of the population, but it was a start.

By this time, the government had begun the process of awarding licences for some of the more rural areas, offering two licences in each of 19 so-called circles or regions. Each bidder had to have at least one foreign partner in its team, as the government had concluded that there was insufficient technical expertise within Indian industry.

Once again there was a problem with the licensing process and the government was obliged to change the rules and re-auction the spectrum. The number of circles any one organisation might own was limited to three, which had the effect of reducing the total licence receipts very substantially. (Difficulties with licensing have been a characteristic of the industry throughout its entire history; allegations of fraud and corruption are commonplace and at least one politician has received a custodial sentence as a consequence.)

Eventually, licences were granted in all 23 circles and the industry began to grow. The number of subscribers more than tripled in 1997, to 827,000; the millionth connection was made at the end of 1998; and by the end of the following year the total had risen to 1.77 million. The 10-million milestone was passed in 2002, and by early 2005 there were 50 million mobile users.

Six companies of real size and stature were emerging from the chaos that was the Indian mobile market: Bharti Airtel, BSNL, Reliance Communications, Hutchison Essar, IDEA Cellular and Tata Communications.

The total number of Indian mobile subscribers had reached 100 million in mid-2006, when mobile penetration hit 10%. This sudden increase was due largely to a change in the licences held by the CDMA operators, which allowed them to offer mobile services to the 30 million customers who were connected to their previously fixed CDMA wireless networks. This ramped up the competitive pressure, and in this environment Bharti seemed to thrive. By the end of 2007 the company accounted for more than 55 million of the country's 235 million mobile subscribers, 15 million more than Vodafone India, the nation's second-largest player. By the end of 2010, when there were nearly 750 million mobile connections in India, Bharti had more than 150 million subscribers, 25 million more than its closest rivals, Vodafone and Reliance.

Bharti had already expanded into a couple of other markets on the sub-continent, Sri Lanka and Bangladesh, and in early 2010 it agreed to acquire most of the African mobile properties of Kuwait's Zain (formerly Mobile Telecommunications Company). This took it into new markets with a population of over 450 million and increased Bharti's subscriber base to over 180 million.

The enlarged Bharti recorded a total profit of $1.3 billion during its 2010/11 financial year on the back of sales of over $13 billion. That was to be its best year – today, it has 279 million mobile customers in India, down from 322 million in 2018, and the company is loss-making. The problem is new entrant Reliance Jio, which has been a disruptive force in the mobile market – arguably the most disruptive force the industry has ever seen.

At the time of writing, in an attempt to cut its debt, Bharti Airtel had proposed to list Bharti Africa, the holding company for its African businesses, on the London Stock Exchange.

British Telecom

This entity dates back to 1878 when it began life as the United Telephone Company. Until the late 1970s it was known as Post Office Telephones, before becoming first British Telecom (BT) in 1981 and later, in September 2001, the BT Group.

Initially, the telephone market in the UK was neither a monopoly nor state owned. Up until 1880 numerous local operators competed for business until a landmark legal ruling determined that a telephone was, in fact, a telegraph, and therefore should come under the control of the Post Office, which was, of course, controlled by the state. The operators weren't happy, so a compromise was reached – the companies could operate as before under the terms of new 31-year licences, after which they would be subsumed into the Post Office.

In 1911 the many different telephone companies in the UK were nationalised and merged, becoming an arm of the Post Office.[87] There began a period that would last for 74 years, characterised by growing inefficiency and restrictive practices on the one hand, and technological invention and brilliance on the other.

By the time of privatisation in 1984, decades of underinvestment and overstaffing had left BT with a near-obsolete network and more than twice the number of employees it needed. Heavily unionised, the company was inflexible and averse to change. Its entire management resources were needed to address its problems, which meant that other issues were neglected, such as the opportunities for growth that existed in markets other than traditional telephony.

For some years after its Cellnet subsidiary was awarded the UK's first mobile licence, BT remained a marginal player in the mobile industry. Then, in 1989, it shocked its supporters and critics alike by acquiring a 22% stake worth $1.2 billion in the American company McCaw Cellular, a pioneer in the field of cellular communications. It was a brave move and really very imaginative, but in 1989 the path to control was blocked by US restrictions on foreign ownership of radio assets.

It's somewhat curious that BT should happily invest in a US cellular operator yet be so averse to opportunities in the same industry much closer to home. Whatever the reason, by 1998 BT held significant stakes in only two European mobile businesses, Cellnet and Cegetel, the controlling shareholder in France's SFR. Apart from these, its sole interests in the international mobile market were a small minority stake in Airtel in Spain and fragments of the equity in the various Tu-Ka and Digital Tu-Ka consortia in Japan.

Then, suddenly, in the mid-1990s, BT started bidding for licences here, there and almost everywhere. By the end of the century it had bought out the Cellnet minority and acquired a 50% stake in Telfort, an alternative network operator that had just been awarded a mobile licence in the Netherlands. It won the fourth mobile licences in Germany and Italy and bought a minority stake in Rogers Canada. Then it turned its attention

---

87 All except that which provided service in the Yorkshire city of Kingston-upon-Hull, which slipped through the net somehow.

to Asia, buying stakes in SmarTone in Hong Kong, Maxis in Malaysia, StarHub in Singapore, LG Telecom in South Korea and another, different Airtel, in India.

Then it all started to go horribly wrong.

Throughout its 2001 fiscal year, cash was flowing out of the company at an uncomfortable rate. In March 2000 it had paid £1,158 million for a 49.5% stake in Esat Telecom in Ireland. The following month it raised this stake to 100% at a cost of an additional £856 million. The UK 3G licence came next, costing another £4,030 million. Three months later in July 2000 Nederlandse Spoorwegen, BT's partner in the Netherlands, sold its stake in Telfort to BT for £1,207 million, which obliged BT to pay the entire cost of a Dutch 3G licence, a mercifully small £266 million.

In the German licence auction in August 2000, VIAG InterKom, BT's 45%-owned associate, acquired one of six licences for £5,160 million. BT's share of that was £2,323 million, which brought the year's investment in mobile assets and licences up to a total of £8.68 billion.

Then Telenor, one of BT's partners in InterKom, opted to exercise a put option, obliging BT to acquire its 10% stake in the business at the cost of a further £1,032 million. In February 2001 VIAG, the other partner, exercised its put over the last 45%, and BT was down another £4,562 million. To this total of £5,594 million was added a further £3,615 million to repay both former partners for their share of the cost of the German licence. The total invested in mobile licences and equity stakes during that phenomenal year amounted to £17,891 million, or just over £49 million every day.

In its 2003 Annual Report, BT congratulated itself for having reduced its net debt from around £30 billion to just under £10 billion – something it achieved by disposing of all the businesses it had patiently built up during the previous decade. Airtel and Cegetel were gone, as were Canada and Japan, and all of the rest of the Asia stuff. Concert's party had also come to an end.

Most significantly, the remains of the mobile portfolio (the UK, German, Dutch and Irish businesses) had been demerged under the rather peculiar name mmO2. There was nothing millimetric about its debt, however – BT had sent it on its way with almost all of the borrowings the group had taken on in the 2001 financial year. O2, as it soon became known, thus began life as an independent entity with a potentially life-threatening £17-billion net debt – equivalent to slightly more than five years' revenues at the then prevailing run-rate. It is to the company's immense credit that it managed to survive.

BT may have got its strategy right when it looked to enter foreign mobile markets, but its timing had been dreadful. It also had to operate in a national market that was far more competitive than those on the other side of the channel – or the Atlantic, for that matter. The UK had been the first country in Europe to introduce any kind of competition in traditional telecommunications, and had also been the first to extend this to every other aspect of the market as well. It had licensed numerous new cable operators to offer telecoms services in addition to television – but BT had been prohibited from reciprocating. Less than 20 years after the first moves to reduce BT's monopoly, the company found itself operating in one of the most competitive mobile markets in the world, with no fewer than five network operators and several early MVNOs.

By 2015 BT was in better shape and returned to the mobile market with the acquisition of Everything Everywhere (EE) for £12.5 billion. There was also the sense that BT was cut a bit more slack by the regulators than O2, Three or Vodafone – early in 2013 it had been allowed to offer 4G services over its 1800MHz spectrum, ahead of the UK's auction of dedicated 4G spectrum (800MHz and 2600MHz), which allowed it to poach many of the most lucrative customers in the UK from its disadvantaged competitors.

Over the last decade BT has had to cope with growing criticism, especially with respect to the level of fibre access. If the company has underinvested in fibre, it might be because it has been unable to earn a sufficient return on its activities: this is the price that has to be paid for two decades of lop-sided, asymmetric regulation.

Thankfully, in the recent past, there has been some loosening of the regulatory reins, and BT is now allowed to provide TV and broadcast services, as well as both fixed and mobile telecoms services.

Cable & Wireless

The 'cable' side of this company dates back to the time of the 'Indian Mutiny' in 1857, following which the British government determined to improve communications with the Empire by creating a network of subsea telegraph cables, to enable a more rapid response should such an event ever happen again. Various privately owned cable companies were licensed for this purpose.

The 'wireless' part was originally Marconi's Wireless Telegraphy Company, the business set up by the Italian inventor.

The two came together in 1929, when the UK government judged the assets were of such fundamental importance that they should be controlled by the state.

That decision was reversed 53 years later by the first Thatcher government, and Cable & Wireless was the first state-owned business to be sold to private investors as part of the wide-ranging reform of the UK's telecommunications industry.

After winning the second fixed-line licence in the UK and acquiring control of Hong Kong Telecom, Cable & Wireless embarked on an extraordinarily successful expansion programme. Its traditional operations in the Caribbean, Hong Kong and elsewhere gave it an enviable international presence, to which it added a long list of mobile investments – in the UK, Pakistan and Germany (1990), Australia, Belarus and Russia (1992), Bulgaria (1993), South Africa and Colombia (1994), France, Germany and Singapore (1995), and Japan (1992–97).

By the mid-1990s Cable & Wireless was describing itself as a 'federation' and talking about linking all these assets via a 'global digital highway' that it had been building since the start of the decade.[88] It wasn't entirely clear what any of this meant in practice. Competitors sniggered about the many changes of tack the company had undertaken, about its 'strategy de jour' approach – but when Mannesmann asked C&W whether it wanted to sell out of D2 in Germany and crystallise a 6p per share increase in earnings, things became a little clearer. It was sitting on a goldmine.

---

88 Cable & Wireless Annual Report 1991.

The value of all the mobile assets had increased spectacularly. They could be sold and the proceeds reinvested in the next great growth market – web-hosting and IP-based services. But as the 1990s drew to a close, in quick succession Cable & Wireless sold everything that had ever made the company money, and bought a bunch of near-worthless, soon-to-become-commodity businesses at astronomical prices, from very willing sellers who could hardly keep a straight face as they signed the contracts.

The various cable-TV business interests had gone and all the wireless assets were gone … One wry observer noted that the only part of the name that was still appropriate was the ampersand. Attempts at creating value from the new business were, broadly speaking, entirely unsuccessful.

Cable & Wireless reorganised itself for one last hurrah, spinning off the relics of its colonial past as Cable & Wireless Worldwide and putting the rest of what was left into Cable & Wireless Communications. Vodafone bought that bit in 2012, and in 2016 Liberty Global swept up the Worldwide group.

China Mobile

The first mobile service in China was launched in Guangdong in 1987 by the Ministry of Posts & Telecommunications. This was part of a concerted effort by the Chinese government to modernise and expand the country's communications, which in turn was an essential element in the multi-faceted programme of industrial modernisation that has transformed the country over the last three decades.

The task was gargantuan. The fixed network in the country was in a similar condition to those in the Soviet Union and Eastern Europe, and was certainly not designed to carry everyday conversations between the country's 1.2 billion people. Only the international gateway worked really well, thanks in part to the generous gift of analogue switching equipment donated by Cable & Wireless from neighbouring Hong Kong.

The targets set by the government looked unattainable: its aim was to increase the number of telephones from one for every 300 people to something nearer three for every hundred by the end of the century. Allowing for population growth, this meant a tenfold increase from the 3.5 million in service at the end of 1986 to something nearer 40 million.

Early progress was painfully slow and the launch of the mobile service did very little to change this; the new network (based on the TACS standard) was very much the preserve of the elite. A year after service had started, there were just 3,227 users on the network. By 1990 the fixed-line base had risen to a relatively modest 6.85 million, while the number of mobile subscribers stood at just 18,319.

Despite this, the government then tripled the target for fixed penetration, with a new goal of 9% (nine phones for every hundred people) by 2000. Remarkably, it met this target, achieving 144 million lines by the millennium, equivalent to 11.5% penetration. Only the US had more telephone lines.

The mobile industry began to make progress too: the number of mobile subscribers more than doubled in 1994, to reach 1.57 million, and more than doubled the following year. By the end of 1997 there were over 13 million mobile users in the country, and at the end of the century just over 85 million.

This astonishing transformation has been very cleverly overseen over many years. As part of the process, the Ministry had established separate Posts and Telecommunications Bureaus (PTBs) for each of the country's 31 regions and created a new holding company for these, the China Telecom Group. Additional capital was then introduced into the businesses through a programme of partial privatisation. China Mobile (Hong Kong) Limited was the result, coming into being in 1997 when the mobile operators in two regions of China were transferred from the regional PTBs to the new company. The following month the company was listed on the Hong Kong Stock Exchange, when 23.5% of the shares in issue were sold to private investors. (More than 20 years later, the Chinese state still owns the vast majority of the shares in the business.)

By this stage China Mobile was not the only operator in the country. In late 1993 the state had made a rather half-hearted attempt to comply with World Trade Organisation obligations, through the awarding of two new mobile licences. These were given to two other organisations that were also wholly owned by the state: China Unicom (China United Telecommunications Corporation) and Great Wall, a company controlled by the People's Liberation Army. (Great Wall's business was subsequently transferred to Unicom and eventually sold to China Telecom in 2008.) Both new networks began operating in 1997, based on the GSM and CDMA standards, respectively.

A few months after China Mobile's foray into capitalism, a new supervisory body was created, called the Ministry of Industry and Information Technology. Over the next six or seven years it supervised the transfer of additional regional operators to the new business.

To maximise the attractions of China Mobile to Western investors, the Ministry of Industry and Information Technology ensured that the first businesses that were given to the new company were those in the most prosperous, economically developed areas, starting with Guangdong, Zhejiang and Jiangsu. This was well received by investors, so more followed in 1999 and 2000, with the operators in a further seven regions passed to China Mobile, including those in Beijing and Shanghai, the country's two largest cities. This deal left China Mobile with just over a third of all mobile subscribers in the country. Further transfers followed, and by the end of 2004 the rest of the regions had come under China Mobile's control.

Over seven years, bit by bit, the government had shifted control of the industry from the state to the private sector – though with 72.72% of China Mobile's own equity still being controlled by state-owned entities, it might be argued that the changes were rather more cosmetic than real. But whatever one may think about this process of privatisation by stealth, there was no doubt that the government had succeeded in its aim.

China Unicom and China Telecom

The same process of creeping privatisation was applied to two other mobile businesses, China Unicom and China Telecom. These were, notionally at least, competitors of China Mobile; a majority of the shares in these former companies were and are also owned by the state.

For most of the recent past the three have operated networks based on different technologies. This suited the state, as it enabled it to experiment with varying technologies, including a home-grown 3G variant called time division synchronous code division multiple access (TD-SCDMA) that was entirely ignored by the rest of the world. Of course, this also made it much more difficult for subscribers to move between networks, as they would have needed to buy a new handset to do so.

The process of privatising the Unicom network began in 2000 when ownership of the GSM networks in 13 regions (including Beijing, Guangdong and Shanghai) was transferred from the Unicom Group to its mobile subsidiary, China Unicom. The new company's IPO saw 22.5% of its shares being sold to private investors and was deemed a success. In early 2003 a further nine provincial networks were transferred, with the remaining nine following 18 months later.

By 2002 Unicom had launched a second network based on CDMA technology that operated in parallel to the GSM network – it seemed that the state was interested in assessing the attractions of both of the world's main standards. This impression was later strengthened when the country awarded 3G licences – China Mobile was given a licence to operate the non-standard TD-SCDMA, Unicom got W-CDMA, while China Telecom drew the short straw and got CDMA-2000 1x EV-DO.

This complex and subtle process had created exactly the situation the state wanted. It now had an ultra-modern communication system without having had to give up control of any of the operators. At the time of writing there are 1.60 billion mobile connections and a further 588 million fixed, making China by far the world's largest telecommunications market.

China has also created one of the world's largest telecoms-equipment manufacturers, Huawei Technologies Company.

Deutsche Telekom

Deutsche Telekom traces its origins back to 1877, just a year after the invention of the telephone. The German postmaster, Heinrich von Stephan, permitted the first trials of the new invention, was impressed with what he saw, and determined that the service should come under the control of the national Post Office. This arrangement prevailed until 1989, when post and telephone services were established as separate commercial entities.

Demand for the new service was considerable, and by the end of the 19th century there were more telephones in service in Berlin than in all of France. The company always placed great importance on technical excellence, and as early as 1908 the first automatic exchanges were introduced. After 1912 it took to burying all new cables in underground ducts. In the 1920s subscriber trunk dialling (STD) was introduced, and an early form of mobile telephony was made available to passengers on the train line between Berlin and Hamburg (the so-called A-net). This was followed in 1933 by the first telex services, a system of telegraphy with printed messages transmitted and received by teleprinters.

In 1933 the Post Office came under the control of the Nazis and the newly renamed Deutsche Reichspost became an instrument of the state. This ultimately led to the destruction of much of the company's property and assets, an all-too-frequent outcome when the state becomes involved with matters of business.

By the end of the Second World War many of the Reichpost's buildings had been destroyed by the Allies' bombing campaign, and many of the staff were dead or missing. The restoration of the country's communications network took a high priority and the business was reconstituted as a state-owned body, with a new name, Deutsche Bundespost.

In 1948 the country's new constitution prohibited private ownership of either post or telecommunications. By the early 1950s the process of rebuilding the network was complete in the newly established Federal Republic of Germany.

In the east, the German Democratic Republic, rebuilding took a while longer, and the process of reconstructing the telephone system there didn't really begin until 1990, when the two parts of the country were eventually reunited. Deutsche Bundespost took over the East German entity, Deutsche Post, which was in severe need of modernisation: there were fewer than 3 million lines for a population of over 16 million people, and more than half of these were served by equipment that was more than 40 years old. An astonishing 2,000 towns and villages were completely unserved, with no access to telephones of any kind. Rebuilding it – the Telekom 2000 programme – was a mammoth task, costing over DM60 billion in all, and it's hugely to the Bundespost's credit that it managed the job as quickly and efficiently as it did.

Changes in EU legislation that proposed opening all the telecoms markets of Europe to competition provided the final impetus, and at the beginning of 1995 the Bundespost was reborn as a public company, Deutsche Telekom AG. Two years later it was privatised when the state sold 26% of its shares in the business to a wide range of private and institutional investors.

By this time, several parts of the telecoms market were already competitive, including equipment supply, paging, some satellite services and, most importantly, mobile communications. The arrival of a second mobile operator galvanised the market, and in its first year, 1992, Mannesmann's D2 network attracted 235,000 subscribers – a total it had taken Deutsche Telekom's analogue C-Net five years to reach. Two years later the new company had nearly a third of the country's mobile users and the overall total had jumped to well over two million.

Deutsche Telekom had in the meantime undertaken a substantial reorganisation. At the same time, it took steps to reduce the potential impact of further domestic competitors, and in the early 1990s Deutsche Telekom Mobilfunk (DeTeMobil), one of several newly formed independent units, acquired interests in four regional franchises in Russia, including one covering Moscow. These eventually would lead to Deutsche Telekom acquiring a 46% shareholding in MTS, one of the three giant mobile operators in the former Soviet Union.

In 1993 Deutsche Telekom formed a joint venture with Ameritech called Magyarcom to acquire an initial 30.1% stake in Matav, the Hungarian state's PTT. Over the next few years, Deutsche Telekom was to provide most of the impetus for rebuilding the fixed network and creating a new mobile system. Ameritech sold out in 2000, leaving Deutsche Telekom with a near-60% stake in Matav and, through that, interests in two other operators that Matav had controlled, in Montenegro and Macedonia.

Deutsche Telekom entered the mobile market in Poland at the end of

1995, when it acquired a 22.5% stake in Polska Telefonia Cyfrowa (PTC), one of two consortia that were awarded licences to operate GSM networks in competition with the PTT's analogue NMT-450 system. Within a few months it had secured another two mobile franchises in the region, in Austria (competing with Telekom Austria) and the Czech Republic (where it faced the Eurotel consortium backed by US West and Bell Atlantic).

In 1995 DeTeMobil bought 25% of Satelindo, a mobile operator in Indonesia, at a cost of $676 million. It followed this in 1996 with 21% of Malaysia's Technology Resources Industries (TRI) for DM900 million and a 34% interest in Islacom in the Philippines for $243 million (10% direct interest and 40% of the main shareholder, Asiacom Philippines). Eventually Deutsche Telekom would dispose of all these investments.

In 1996 Deutsche Telekom and France Telecom joined forces to create Atlas, a 50/50 joint venture to provide telecoms services to international businesses. This then formed an alliance with Sprint, called Global One. Both European telcos acquired 10% interests in the US operator. This arrangement was not destined to last either: there were disagreements about the budget and, specifically, the level of investment that was needed, and in January 1997 France Telecom bought out its partners.

A second joint venture followed in late 1997, again with France Telecom, but this time including the Italian power company Enel, which owned a 51% stake. This didn't fare much better. The consortium was successful in its bid, and was awarded a licence in early 1998 to operate both fixed and mobile networks. The following year brought trouble, however, when France Telecom alleged that Deutsche Telekom had breached the shareholders' agreement by discussing a possible merger or joint venture with Telecom Italia. It took the matter to the International Chamber of Commerce for arbitration and by 2000 Deutsche Telekom had agreed to sell its shares in the venture.

Deutsche Telekom also became the partner of choice for several Eastern European telecoms companies, taking stakes in Croatia's Hrvatske Telekomunikacije in 1999 and Slovak Telekom in 2000.

In late 1999 Deutsche Telekom re-entered the UK mobile market, buying one2one from Cable & Wireless and MediaOne. One2one was the first of the UK's PCN operators, but it had been outmanoeuvred by Orange and was in fourth place in the UK market. However, at just £8.4 billion ($13.36 billion) it cost a lot less than the amount Mannesmann would soon pay for Orange. The £4 billion it would have to pay for a 3G licence in 2000 changed the mathematics somewhat, but eventually Deutsche Telekom managed to earn a reasonable return.

This gave the company a taste for adventure and what came next put a severe strain on the German giant's finances. In 2001 Deutsche Telekom acquired VoiceStream, the US company backed by Hutchison, for $33 billion. However, the total cost was nearer to $50 billion including the $5 billion spent on extra spectrum licences, a similar amount for the assumption of VoiceStream's debt, and a further $7 billion for the acquisition of PowerTel, another GSM-based company that had a complementary footprint to that of VoiceStream. The best part of 20 years on, this vast investment is finally beginning to generate an appropriate return.

By the end of 2002 Deutsche Telekom had completed the acquisition of the Dutch company Ben (now T-Mobile Netherlands), and had become the world's second-largest multinational mobile operator, after Vodafone. Its home market was the fourth largest in the world. Outside this, it controlled businesses in seven other markets, including the UK and the USA. It also had investments in four other smaller operators, which extended its reach into a large part of eastern Europe.

It was pretty well positioned, but it was time to tidy up the portfolio. The Asian mobile assets were sold: Satelindo went in 2002, the other two followed a year later. Assets where there was no likelihood of control were also pruned, including the 16.3% shareholding in Ukrainian Mobile Communications, which was sold to MTS. Deutsche Telekom's interest in MTS was reduced in 2002 and again in 2003, before a final disposal in 2004. Shares in the GEO satellite company SES Global were sold in the same year.

Deutsche Telekom then went back on the offensive. In 2006 it acquired tele.ring, the number-four mobile company in Austria, and raised its stake in PTC in Poland to 49%. It took a 25% stake in OTE, the Greek national telephone company, in 2008, increasing this to 40% in 2011. OTE had its own international ventures, so as well as Greece, the deal gave Deutsche

Telekom a presence in three other Eastern European markets – Albania, Bulgaria and Romania.

In 2010 it merged T-Mobile UK with Orange UK, to create Everything Everywhere (EE).

Three years later it strengthened its presence in the US with the acquisition of MetroPCS, one of the more successful independents, for around $11 billion.

Most recently, BT acquired EE for £12.5 billion in a deal that took Deutsche Telekom out of the UK market once again, while bringing BT back in. The sum paid by BT is almost exactly what Deutsche Telekom paid for one2one and its 3G licence (£8.4 billion and £4 billion, respectively) but, of course, one should also take into account the £31 billion France Telecom (now Orange) paid for Orange UK and its three small European associates. This is a rather better example of 'destruction of shareholder value' than Vodafone's much-criticised acquisition of Mannesmann.

Today Deutsche Telekom has a strong position, both at home and in international markets. However, it spent rather a lot of money to achieve this, and net debt is still close to €60 billion and debt to equity a rather uncomfortable 135%.

(CK) Hutchison

The Hong Kong-based conglomerate CK Hutchison (formerly Hutchison Whampoa) has been a significant player in the telecoms market for over 30 years, first becoming involved in the industry through a mobile licence in its home territory back in 1985.

At that time, the business of telecommunications in Hong Kong was dominated by Cable & Wireless, which had a monopoly of both local and international call traffic. It was an open secret that Hutchison would have liked to own Cable & Wireless's largest subsidiary, Hong Kong Telecom, and at one stage in the late 1980s it even began amassing a stake in Cable & Wireless ahead of a possible takeover bid. That never materialised, but Hutchison compensated by acquiring interests in other markets as and when suitable opportunities arose.

After entering the mobile market in Hong Kong, Hutchison's next move was to acquire one of the four telepoint operators that had been licensed to offer a limited-range, quasi-mobile service in the UK. That proved to be an expensive failure, but the company was soon back in business, taking a stake in Microtel (later Orange). That proved to be a spectacular success and in late 1999 Hutchison sold its interests in the business to Mannesmann, triggering a sequence of events that would lead to Vodafone's acquisition of Mannesmann.

In 1992 Hutchison made an investment in an Indian partner to create Hutchison Max Telecom, which won a licence to operate a mobile service in Mumbai, one of the first four regions to be licensed. This business expanded rapidly in the early 2000s through the award of additional regional licences and the acquisition of smaller competitors until it had achieved full national coverage – at which point, wily Hutchison sold the business to Vodafone.

Throughout all this, Hutchison had moved into several new markets. In 1997 it formed the Partner Communications consortium to bid for the third mobile licence in Israel. It began operating in 1999, was listed on the Tel Aviv Stock Exchange and was eventually sold to the Scailex Corporation in 2009.

In 1997 Hutchison acquired a 19.9% stake in a new company, VoiceStream PCS, a subsidiary of Western Wireless, an American company specialising in rural cellular markets that had been formed by three former McCaw Cellular executives. VoiceStream had been established to acquire licences in the PCS auctions and Hutchison's investment – $248 million in the first instance and a further $1 billion two years later – allowed the business to acquire several other US PCS operators during the course of 1999, to create a company with a nationwide GSM presence. Two years later this business was sold to Deutsche Telekom at a price of over $30 billion.

In 1998 Hutchison moved into Ghana, launching a CDMA network in the Accra/Tema region. This it sold in 2008 to EGH International for $584 million.

In 1993 Hutchison launched a mobile system in Sri Lanka. This grew slowly at first, hampered by the lack of a presence in the capital, Colombo, but eventually obtained a full national licence after which it merged with Emirates Telecommunications' (Etisalat's) Sri Lankan business in 2018.

Hutchison acquired a fourth mobile licence for a GSM system in Paraguay in 2000, selling this to América Móvil five years later.

In 2003 it began marketing a CDMA service in Thailand. This was a joint venture with CAT Telecom, the former communications authority of Thailand. A similar arrangement was made the following year in Vietnam; this also started as a CDMA network, but the business was hampered by a shortage of appropriate handsets and eventually managed to switch to GSM.

In 2005 Hutchison acquired Charoen Pokphand's 60% interest in the Indonesian company PT Cyber Access, renaming this Hutchison CP Telecommunications. A reorganisation of Hutchison's telecoms assets brought these last three together as Hutchison Asia Telecommunications, a company that had just under 50 million customers at the end of 2018.

Hutchison 3 Group Europe

Hutchison set up the 3G Group as a separate business after it had acquired the A-block licence in the UK 3G auctions. In the weeks that followed, Hutch bought out its partner in the auction, the Canadian company TIW, and formed a new alliance with NTT DoCoMo and KPN. It sold both interests in the licence for a total of about £2.1 billion, thereby recouping nearly half of the £4,385 billion it had paid.

Over the next year it acquired five more European licences: it formed a consortium in Italy with Tiscali, which won a 3G licence at a cost of €3,254 million; in Austria it paid €113.6 million for two spectrum blocks; and it formed a 60:40 joint venture with the Swedish company Investor AB, which acquired licences in Sweden and Denmark. A sixth and final licence was acquired in Ireland.

In the years that followed, Hutchison increased its holdings in all of these ventures. In particular it bought back the stakes in 3UK from DoCoMo and KPN at a deep discount – £210 million, a tenth of the amount for which they'd been sold just three years earlier.

In 2015 the renamed 3 Group announced that it intended to merge

3 Italia with Wind, to form a new market leader in Italy. The deal was initially blocked by the European Commission, but after the two companies agreed to sell some of their spectrum to facilitate the entry of a new player, the Commission relented. The deal was finalised in 2016.

Just over a year later, Wind's owner VEON announced that it was selling its stake in the joint venture, leaving Hutchison as the sole owner of Wind Tre. VEON received a payment of €2.45 billion for its share of the venture.

KDDI Corporation

KDDI, the number-two operator in Japan, was formed through the merger of two cellular operators, Nippon Idou Tsushin (IDO) and Daini Denden (DDI), with Kokusai Denshin Denwa (KDD), a company that had been formed in 1953 to operate international traffic out of Japan. Both cellular operators were owned by companies that had recently entered Japan's long-distance carriers as competitors to NTT.

The strategies and ambitions of the two varied dramatically, however.

DDI was the brainchild of Kazuo Inamori, the chairman of Kyocera Corporation, which had begun life as a ceramics company, but had diversified into various other businesses that today we might call IT-related stuff – computers, digital switches and the like. In 1983 Inamori noticed what was happening in the US telecoms market and thought the same kind of liberalisation might be possible in Japan. He saw NTT as distinctly analogous to AT&T, only worse – its long-distance business was characterised by high prices and poor service. DDI's opportunity was clear.

Others weren't so sure and scepticism about the project increased when two other companies announced plans to enter the same long-distance market. Both seemed to have an advantage over DDI. Japan Telecom had a typically complicated ownership structure but, crucially, its shareholders included companies from the Japanese Railways Group, which were immediately able to provide access to many thousands of miles of very valuable rights of way alongside railway tracks on which to lay fibre cables. And Teleway Japan had been formed by the Toyota Motor Company in conjunction with the Japan Highway Corporation, so it shared a similar advantage – thousands of miles of motorway.

DDI couldn't match that, but Inamori had seen what MCI's Bill McGowan had done in similar circumstances, so he opted for a combination of long-distance microwave links backed up by a deal to obtain access to some of NTT's network on a wholesale basis. DDI had to calm the fears of environmental and residential pressure groups protesting about the dangers of radiation, but by 1987 it had built and launched its long-distance network, put a strong management team in place, and begun to establish a reputation for quality as well as value for money. DDI was beginning to take market share away from NTT and all looked well.

But Inamori noticed that pressure on pricing was slowly destroying the profitability of the fixed telephone business, and started looking for an encore to his long-distance activities. He lighted on the mobile-phone business, and applied for mobile licences in the various Japanese regions. The government had reserved two of the most valuable for another consortium led by Teleway Japan, but DDI was awarded the right to operate in several parts of the country, including Kansai, a major industrial centre.

Looking to differentiate the new business's offering from that of NTT, DDI opted for the 'foreign' (European) TACS standard, and further upset many of the more conservative elements in Japanese society when he announced that DDI would be using equipment provided by Motorola, because it was better and cheaper than Japanese equipment.

During the course of 1988, ahead of the launch of the new service, DDI looked for a partner to share the load and chose the regional electrical power companies. With their support, the company was awarded four further regional licences in 1989, in the regions of Tohoku, Hokuriku, Hokkaido and Shikoku. The last unserved region, Okinawa, was to follow in 1991.

The two reserved regional licences were given to Teleway Japan's new vehicle, IDO. The first of these served the Kanto region, an area that contained about a third of the country's population and included the whole of Greater Tokyo. The second was for the Tokai region, a heavily industrialised area to the southwest of Tokyo. Together, these gave IDO access to more than 40% of the country's population of 125 million.

IDO was to prove less of a success than its competitor, especially in the earliest years. Although it was the first to launch, in 1988, and had a far smaller geographic area to cover, it was hampered to a degree by the demands of Toyota, the main shareholder in Teleway, which required it to focus on car phones rather than the newer hand-portables.

DDI began operating in 1989, in Kansai, the largest region available to it. It concentrated on smaller hand-portable devices, and by the end of 1992 it had completed its national network and negotiated a national roaming agreement with IDO. It acquired nearly 350,000 subscribers, giving it a 42% share of the market in its eight regions, while IDO's base of 288,000 equated to 36% of its available market.

Both operators then introduced Japan's new Personal Handyphone service – a short-range quasi-cellular system, not hugely different from the UK's CT-2 networks. These were far more attractively priced than traditional cellular phones – both to buy and to use – and although they were restricted in their utility, they found a market niche as they enabled Japanese youths to walk far enough away from their parents' homes to avoid being overheard.

In October 2000 DDI merged its long-distance business with KDD and IDO. The three-way merger created KDDI. The following month, DDI merged its regional mobile businesses to create the au Group. This merged with KDDI in October 2001.

Since then, although the enlarged business has grown substantially, it has lost ground to both DoCoMo and J-Phone. Technology choices played a part here. Its decision to adopt America's CDMA in preference to the Japanese PDC standard was a mistake – even though CDMA had some advantages over PDC, the network equipment was more expensive and the handsets a lot less attractive. It also gave rise to interoperability issues following the merger, as the more traditional IDO had gone for PDC. KDDI's market share suffered as a result.

KPN

Koninklijke KPN (Royal KPN), the national telephone company in the Netherlands, has its origins in the 19th century. In the first years following the invention of the telegraph and the telephone, demand for these services was modest, so in 1886 the Netherlands government took the decision to bring the two together within the state-controlled postal system, thus forming one of the world's first PTTs. The organisation was given a surprising amount of freedom, having the ability to set its own budgets and make investment plans without having to refer to the government every step of the way.

Demand for telecoms services didn't really become established until the late 1940s. By then the German occupation of the Netherlands had seen the PTT lose its independence and much of the company's plant and equipment had been destroyed. The postal services were also suffering – postmen, like everyone else in the Netherlands, complained that they had lost their bicycles following the German withdrawal, so the telephone service took up much of the slack, becoming the dominant element within the business.

In 1989 the telephone company was reorganised and became a private business, adopting the name PTT Nederland. The state was still the sole owner of the business, and to many outsiders, PTT Telecom looked more like an arm of the government than a commercial enterprise. That perception began to change after the Kingdom of the Netherlands sold a 30% interest in the renamed KPN in 1994, and especially after a further disposal in 1995 had reduced its holding to 45%. This made the company a rather more attractive proposition.

Although the network had been greatly expanded and extensively modernised, KPN was small by the standards of the industry. The proposed deregulation in European telecoms services was clearly a threat to its traditional business market, especially the lucrative international side of that service, so KPN looked for a partner to help increase its scale and scope. It found one in the shape of Televerket (now Telia) and, shortly thereafter, the Swiss PTT. These three joined forces to address the international business market under the banner Unisource. Telefónica later joined the consortium.

Meanwhile, the first mobile services had been launched in the Netherlands. An analogue NMT-450 system began operating in 1985 but within a couple of years this was becoming capacity constrained, so a second analogue network based on the same NMT technology, but at the higher 900MHz frequency, began operating in 1989.

KPN enjoyed a monopoly of mobile services in its home market for more than a decade, not facing competition until Vodafone's Libertel consortium opened its GSM network for business in 1995.

Recognising that it was unlikely to win any of the highest-profile competitions for mobile franchises, KPN focused its international expansion efforts on Eastern Europe. As early as 1991, it had taken a 10% stake in Ukrainian Telecom, the national PTT. It followed this a year later, forming a consortium (with Deutsche Telekom and TDC) that acquired a 49% interest in Ukrainian Mobile Communications, the country's first mobile network.

Two years later KPN joined forces with TDC once more, this time in conjunction with the other three large Scandinavian telcos, Telia, Telenor and Telecom Finland. Three local investors were recruited and the team was subsequently awarded a GSM licence in Hungary – the first instance of competition in mobile in the region.

Back at home, having gone to the expense of overlaying a 900MHz network on top of the 450MHz one, KPN was looking to squeeze as much revenue out of the old system as it could and didn't put much effort into selling GSM. It eventually launched its first digital system in 1994, two years after most other European operators had launched GSM. This was a mistake: within a year, Libertel had taken 28% of the market for digital mobile, equivalent to 18% of the total at that stage. Within three years it had 35% of the total market.

KPN made additional investments in Telkomsel in Indonesia, Eircom in Ireland, Cesky Telekom in the Czech Republic and a number of smaller properties, giving the company quite an impressive international presence, but trouble was just around the corner.

KPN had formed a loose alliance with Hutchison and DoCoMo to cooperate in the European mobile market and, with this in mind, acquired a 77.49% stake in E-Plus in Germany in late 1999, from BellSouth (which retained a 22.51% stake and shared control of the business). E-Plus doubled KPN's presence in the mobile market, but at over €20 billion it was a very big deal and it put considerable strain on the comparatively modest balance sheet, which was exacerbated by the events of the following year.

KPN had planned to use the 3G auctions as a way to create a pan-European platform in partnership with the two Asian companies. The strategy began well enough, with KPN signifying its good intentions towards its partners by taking a 15% share of Hutchison's new 3G licence in the UK at a cost of €1.5 billion in 2000. This was followed by an easy win at home in July that year, where KPN's licence was acquired for a modest €711 million.

Next came Germany. KPN had joined forces with Hutchison to bid for one of the available licences. The idea was that the two companies would share the cost of the spectrum, but after the price rose to €8.4 billion for just two 5MHz blocks of paired spectrum, Hutchison decided that the two blocks were insufficient for what it had in mind, so it abandoned the partnership. KPN was left to fund the whole of its €6.5-billion share of the licence fee on its own.

With a balance sheet showing debt of €22 billion and a debt-to-equity ratio of 160%, KPN was forced to take drastic action. The sale of a 15% stake in KPN Mobile to DoCoMo raised €4 billion, then, one by one, KPN liquidated its interests in the Czech Republic, Hungary, Indonesia, Ireland, Ukraine and the USA, raising a further €3 billion. It was left with just three main markets, Germany, the Netherlands and Belgium, plus a lot of broken dreams.

KPN limped back in 2005, acquiring Telfort from the equally damaged BT and merging this with its own KPN Mobile.

Then, in 2012, the unwanted attentions of América Móvil awoke KPN from a seven-year slumber and, to help repel the unwanted approach, it tried to sell its German subsidiary E-Plus to Telefónica. But the Spanish company had its own problem – a €62-billion debt mountain and a 235% debt-to-equity ratio. In the end, KPN had to deploy a 'poison pill' by issuing special shares to the KPN Trust, which effectively stymied the Mexican company. Some were left wondering why they'd bothered.

The Spanish came back to the table in 2014 and offered to merge KPN's German business with its own German operation, offering KPN €5 billion in cash (net of debt) plus a 20.5% minority stake in an enlarged Telefónica

Deutschland business. It wasn't a great price, but KPN didn't have a lot of choice. The deal went through, reducing the number of mobile networks in Germany from four to three.

Finally, in 2015, KPN agreed to sell Base, the Belgian mobile business it had formed with (the original) Orange back in 1988. The buyer was Telenet, a Belgian cable company, which had previously operated as an MVNO. It paid KPN €1,325 million for the business, which at the time had just over 2.8 million customers.

The deal brought KPN back to the place it had been more than two decades earlier: it was now a rather dull mid-rank European telco.

MCI WorldCom

MCI (Microwave Communications Inc.) played a pivotal role in the history of late-20th-century telecommunications.

Formed in October 1963, MCI applied for licences to build a series of microwave relay stations between Chicago and St Louis, which would allow it to provide long-distance connectivity to users of two-way radios. Under the leadership of William (Bill) McGowan, MCI grew rapidly. By 1972 it had become a public company and, as far as AT&T was concerned, public enemy number one.

MCI began lobbying politicians and the FCC, insisting that AT&T was violating federal antitrust (anti-monopoly) laws, and when in 1974 one part of the Bell System refused to provide an interconnection agreement on reasonable terms, McGowan filed an antitrust suit against AT&T.

Although it took several years, MCI eventually won. The suit was the catalyst for the dismemberment of the original Bell System and was hailed as a triumph for the spirit of free enterprise. MCI went on to become the second-largest player in the US long-distance market, behind AT&T.

The arrival of cellular technology in the early 1980s provided MCI with a further opportunity to attack the old order. The company formed a new division, MCI Airsignal, to own and operate non-wireline cellular networks, and acquired licences in several key US markets.

In 1995 McGowan changed his mind as to how best address the mobile opportunity. He sold the cellular assets to another entrepreneurial upstart, Craig McCaw, and bought Network Cellular Service of Valley Stream, New York, at the time the largest cellular reseller in the US. The $190-million price tag swallowed up all of the $120 million in proceeds received from McCaw for Airsignal.

MCI believed that with 120MHz of new PCS spectrum about to be auctioned, there would be no shortage of capacity in the mobile market. McGowan hoped to strike deals with cellular operators that would allow him to bypass the local exchange companies and provide a service that integrated fixed and mobile offerings, without the need for expensive investment in licences. The logic was sound enough, but the idea didn't really fly, and by the end of 1997 MCI had been acquired by an even more aggressive, entrepreneurial company, the now notorious WorldCom.

WorldCom started life as Long Distance Discount Service (LDDS) in 1983. The company's first decade was unremarkable, but then Bernie Ebbers, the chief executive, met Jack Grubman, a leading Wall Street analyst and former AT&T employee. Grubman persuaded Ebbers to buy WilTel in 1995, and the company changed its name to WorldCom. The following year it bought Metropolitan Fiber Systems (MFS) Communications in a $12-billion stock swap. It topped that in 1997 with the acquisition of MCI, for which Ebbers paid over $30 billion in WorldCom stock, assuming a further $5 billion in debt.

The following year, WorldCom tried to crash the party when it bid for AirTouch. The shareholders didn't like it and, for once, Grubman failed to deliver. On the rebound, it immediately bid for Sprint, but that initiative also failed.

After that, it all went horribly wrong. Ebbers had borrowed vast amounts to further some of his extracurricular activities, debts that were secured against his holdings in MCI WorldCom stock. As the value of those shares began to plummet, his bankers became increasingly nervous about their exposure. He hit on the idea of creative accounting in an attempt to revive the stock price. In 2002 WorldCom admitted that it had misrepresented its earnings, inflating them by $3.8 billion – the largest such fraud in history.

In 2005 Ebbers and his finance director, Scott Sullivan, stood trial. Both were found guilty. Sullivan testified against his former boss and received a five-year stretch, while Ebbers is now most of the way through his 25-year sentence.

Verizon bought the business in 2005, after it had emerged from bankruptcy protection, an outcome that would no doubt have appalled Bill McGowan had he been alive to see it.

Millicom International Cellular

It's probably safe to say that Millicom was the first to see the potential in the international mobile market, when it joined up with Racal in 1982 to bid for the second licence in the UK.

It sold that stake in 1986 but continued acquiring licences and franchises in subsequent years. Most of these were in smaller emerging markets, but it also recorded a few successes in countries with greater potential, including Chile, Hong Kong, Pakistan and the Philippines. It was one of the first companies to invest in Russian franchises, winning the licence for the region surrounding Moscow in 1990. It was also an early investor in India.

In 1990 Millicom became Millicom International Cellular, through a merger with the Swedish company Kinnevik – the main shareholder in that country's Comvik network. By this time, it had sold several of the stakes it had acquired in the 1980s, including Chile and Hong Kong, but it still had an impressive presence. In Latin America it was active in Bolivia, El Salvador, Guatemala, Paraguay and the Mexican state of Monterrey. In Africa it had interests in Ghana and Mauritius. Its Asian operations included Cambodia and Sri Lanka, as well as Pakistan and the Philippines. And in Europe it had an interest in a Lithuanian operator.

During the rest of the decade, it expanded its interests on all four of these continents. Vietnam, Cambodia and Laos were added to its Asian business. There were five new licences in Africa – in Tanzania, Senegal, Sierra Leone, the DRC and Chad. In Latin America it added Honduras and one of the larger markets in the region, Colombia. Finally, there were new properties in the former Soviet Union, in Estonia, Kazakhstan and Russia.

Millicom then began to narrow its focus. In 2001 it sold its Russian assets to its sister company, Tele2. It then disposed of all its Asian businesses, leaving it with two clear but distinct geographic areas, Africa and Latin America.

In 2016 it reduced its exposure to Africa, selling its business in the DRC to Orange. The following year it merged its Ghanaian businesses with that of Bharti Airtel, and sold its network in Senegal to a consortium that included NJJ Capital, a company owned by Xavier Niel, a French billionaire.

In 2018 it closed another deal with Bharti, when the Indian company acquired Millicom's network in Rwanda.

Most recently, it sold its business in Chad to Maroc Telecom, leaving it with only one remaining African company, in Tanzania.

It offset these sales with acquisitions of cable-TV businesses in Latin America and is looking to offer converged services in several markets there.

NTT DoCoMo

NTT DoCoMo is the largest mobile operator in Japan and is a subsidiary of the giant Nippon Telegraph and Telephone (NTT) Corporation, the former monopoly operator.

It was the first company in the world to launch a cellular mobile-phone system in late 1979, but having done so, it did nothing to promote or encourage its use. The service was so expensive that it was well beyond the reach of most consumers, and indeed most businessmen, and the price of renting a handset was also very high (up until 1993 it wasn't possible to buy a handset outright in Japan). At the end of its first decade the company had just 500,000 subscribers, or about 0.4% of Japan's population.

Although it enjoys a commanding position in its home market today, it has never established any kind of meaningful international presence. NTT DoCoMo has dabbled with minority stakes in various companies over the years, including AT&T Wireless and a couple of Hutchison's ventures, but it never committed itself much beyond this.

Orange/France Telecom

The companies that ultimately formed France Telecom came into existence just a few months after Alexander Graham Bell invented the telephone. Following a series of mergers, the industry consolidated to create the Société Générale des Téléphones. This was nationalised in 1889 and became an arm of government, until 1993, when it was reborn as a commercial enterprise once again, with the name France Télécom. The accents were dropped and later, following the acquisition of Orange plc, France Telecom became Orange SA.

Although France was an early adopter of the phone, the new invention wasn't especially well received there, and shortly before the outbreak of the First World War there were just six phones for every thousand inhabitants, far below the 21 per thousand in neighbouring Germany – not to mention the 98 per thousand in the USA.

Although more modern instruments were introduced, along with the first automatic exchanges, France still lagged as the continent prepared for another world war. Penetration stood at just 3.8% compared to the UK's 6.7% and America's 15.3%. Most telephones in France at this stage were in the hands of businessmen or those of the upper strata of society; and most were in the urban north (and especially Paris) or the holiday towns of the south coast.

After the Germans left in 1945 the phone system needed rebuilding, of course. Work began on digital switching but was then abandoned. Experiments were undertaken with early fibre-optic cables. Space was looked into. Finally, in the early 1960s, a coordinated effort was made to modernise, with digital switching introduced across the whole PSTN network, and by the early 1990s France had one of the most modern phone systems in the world.

France's mobile system had an equally inauspicious beginning. The country's first network was launched by France Telecom in November 1985 and was met with almost complete indifference by an unenthusiastic public – by the end of the year there were just 100 connections. The problem, in large part, stemmed from France Telecom's choice of technology: rather than opting for any of the well-established analogue standards – AMPS,

NMT or even TACS – the company selected a technology called RC-2000, which had been developed by Matra, a French defence-electronics company which was looking to diversify.

France Telecom remained an arm of the French government until 1993, when it changed its status to that of a *société anonyme* (the equivalent of a public limited company, or plc). While the networks might have been modern, the organisation was not – it was still very much a state-owned monopoly, averse to the concept of competition and appalled by the idea of privatisation, but over time it became somewhat more pragmatic. It invested in numerous international ventures in almost every aspect of telecommunications. It didn't want competition at home, but was happy to buy in to new entrants; it didn't want to be privatised, but had no problem investing in businesses that were.

By the end of the 1990s, France Telecom had stakes in fixed-line and internet businesses, in tower and transmission companies as well as mobile. Its mobile portfolio consisted of stakes in ten European operators including Wind, Italy's third mobile operator, Panafon in Greece and Mobistar in Belgium. Farther afield, outside Europe, it owned shares in five mobile operators, including 10% of Sprint, the third-largest American network.

Looking beyond mobile, it had invested in several privatisations, including those of TPSA in Poland, where it had a 24.5% stake, Telecom Argentina (19.5%) and Teléfonos de México (5%). It owned part of Britain's ntl cable operator (formerly National Transcommunications Ltd and later Virgin Media). It had a stake in Crown Castle, the tower company, and also, in Deutsche Telekom.

In 2000, Vodafone's investment-banker friends persuaded France Telecom to part with £25.1 million to acquire Orange, the UK's number-three mobile operator. Adding the cost of its 3G licence (£4,095 million) and its debt (£1.8 billion) brought the total to £31 billion. The acquisition gave France Telecom another 10 million connections – but also transformed the company's balance sheet. A few numbers make the point clearly enough: at the end of 1999, France Telecom had had net debt of just under €11 billion, most of which was long term. A year later this had risen to €61 billion, more than €16 billion of which was due to be repaid within

12 months. The level of goodwill – the premium paid for assets compared to their book value – had leapt from €1.2 billion to €36 billion, while other intangible assets had also increased sharply, from €0.9 billion to €16.3 billion. At the end of 2018 the company now known as Orange was valued at €38 billion, €8 billion less than it had invested to acquire the UK company.

Orange's acquisition of Orange (as it were) was the clear high-water mark in its merger-and-acquisition history: nothing it has bought since has made a material difference to the business. Franchises were added in international markets, most notably Africa, but these were more like notches on a belt rather than real, meaningful investments. And the original Orange business in the UK was itself sold.

The only thing the French company has to show for its €31-billion punt is a six letter name – that's over €5 billion for each letter.

Propel

At the end of the 1990s Propel was one of the largest owners of mobile assets in the world, with (mainly non-controlling) interests in numerous businesses. In Latin America, Propel had stakes in mobile companies in Argentina, Brazil, Chile, the Dominican Republic, Uruguay and in five of the nine regional operators in Mexico. It also owned shares in three Middle Eastern operators, MobiNil in Egypt, Pelephone in Israel and Fastlink in Jordan. Elsewhere, there were investments in Omnitel in Lithuania, Bakcell in Azerbaijan, Mobilink in Pakistan and Hutchison's cellular business in Hong Kong.

The business is better known as having been the Network Management Group of Motorola, which at the time was one of the leading suppliers of both handsets and network equipment to the industry. Motorola supplied all the network equipment for the businesses listed here and had also taken equity stakes.

By 1999 the arrangement was becoming too complicated and conflicts of interest were becoming more commonplace and less easy to resolve, so the decision was taken to cut the business loose. The company prepared a prospectus in 1999, ahead of a $500-million listing on Nasdaq – which never took place. Almost all the controlling shareholders in the various businesses exercised call options to acquire Propel's share of their businesses, leaving it with next to nothing to sell.

SoftBank Group

SoftBank was founded in 1981, with a lofty ambition: it aimed to be 'the world's leading infrastructure provider in the information industry'.[89] The company was initially a distributor of software for use on PCs, a market that at the time was in its infancy.

SoftBank entered the telecoms market when it acquired Japan Telecom, the fixed-line business that Vodafone had sold to private equity in 2003.

Two years later SoftBank acquired Vodafone KK, the third mobile operator in the country. The business had huge potential, but Vodafone saw it as underperforming and hadn't managed to find a way to change that. SoftBank achieved a transformation in a few short months, however, making this a brilliant deal done at a bargain price. The acquisition propelled SoftBank into the top tier of the world's mobile operators.

In 2012 the company bought an 80% stake in another number-three operator – Sprint Nextel, in the US market. SoftBank's CEO, Masayoshi Son, was confident that the experience he'd gained from transforming Vodafone KK could be brought to bear with Sprint, but so far this hasn't happened.

Sprint Corporation

Sprint's origins go back in part to the Brown Telephone Company, which was founded in 1899 in Kansas. It began as an independent local telephone operator but almost immediately entered the long-distance business. Brown grew both organically and through acquisition. By the 1960s it was operating local and long-distance service in 15 states under the name United Utilities. By the middle of that decade it was beginning to experiment with

---

89 SoftBank Group website.

fibre-based networks, which it believed were preferable to the microwave technology used by its principal competitors.

Sprint's other ancestor was the Southern Pacific Railroad Company, one of the earliest US railway operators. In the late 1800s Southern Pacific used telegraph, and later telephones, which lay alongside the many miles of tracks it owned. It decided to sell access to the network, and by the 1970s it was beginning to make inroads into AT&T's long-distance market. It rebranded, taking the name Southern Pacific Railroad Internal Network Telecommunications (Sprint).

In 1983 the long-distance business was acquired by GTE, the largest independent operator in the US. This subsequently merged with US Telecom in 1986 (a subsidiary of the American United Telecommunications), becoming US Sprint Communications. GTE sold its share in the business back to United Telecommunications in 1990.

By this time the company had become involved in the cellular market. It had acquired Centel Corporation, one of the larger independent cellular operators, in 1993, but was still essentially a regional operator. Sprint realised it needed something on a far greater scale if it were ever to be established on the national stage.

The 1995 PCS auctions provided it with the opportunity, and Sprint formed a consortium called Wireless Co. with three cable-TV companies, TCI, Comcast and Cox, and managed to acquire licences for 29 of the 50 MTAs at a cost of $2.1 billion. This gave it a near-national presence, with franchises in New York and Los Angeles, and another six of the top ten markets, but missing out in Chicago and Houston.

Wireless Co.'s shareholders had plans to offer a combined (multiplay) service that encompassed mobile, internet, entertainment and voice, including long distance. The partnership started well enough, and by the end of 1997 it had launched service in 150 metropolitan areas and connected 900,000 subscribers. But cracks in the façade were appearing. The three cable companies jumped ship in 1998, leaving Sprint to go it alone.

It did, for the best part of seven more years, building up the business sufficiently to become the clear number three in the US market, behind

Verizon and the merged SBC/AT&T Wireless. Then, in 2005, it merged with Nextel.

Nextel started life as Fleet Call, a radio-dispatch business using special-ised mobile radio (SMR) technology. Shortly after the launch of the original analogue cellular service in the US, Fleet Call's management realised that new digital versions of SMR could be used to provide 'mobile telephone services, two-way dispatch, paging and alphanumeric short-messaging services using a single, multi-function subscriber unit' (and, in the future, 'data transmission capabilities').[90]

Fleet Call merged its traditional operations with a subsidiary of Motorola (the developer and manufacturer of the iDEN technology used in the system) and emerged as Nextel. As it had already acquired many unloved dispatch businesses across the country, it started out with an almost complete national network. It thought it might be able to steal a march on the cellular operators by offering a national service at a single rate, eliminating all roaming charges. This gave it something of an edge and, typically, its users generated higher-than-average monthly revenues … but not for long. Soon enough, the other US networks began offering similar packages.

The limited choice of iDEN handsets also began to have an effect on the business, particularly after the arrival of the first iPhones. Nextel's customer base started to be eroded, slowly at first and then at an accelerating pace, until eventually, in the second quarter of 2013, the final iDEN subscriber disconnected and the network was no more.

Sprint has struggled on ever since. The alliance with SoftBank provided new capital, but those who expected a miraculous renaissance have been disappointed. In 2018, T-Mobile USA announced that the two businesses planned to merge to create a much stronger number three in the American market, behind AT&T and Verizon. At the time of writing the deal appears to have received conditional approval from the regulators.

---

90 Fleet Call 1996 10-K, filed with the SEC on 31 March 1997.

Telecom Italia

The first telephone companies in Italy date back to the 1880s. The market was served by four separate operators, each of which had its own geographic region. These were brought together in 1925 to form the Società Italiana per l'Esercizio Telefonico (SIP). This was in turn, in the 1930s, brought under the control of the Società Torinese per l'Esercizio Telefonico (STET) and, later, the Società Finanziaria Telefónica, a holding company controlled by the state Istituto per la Ricostruzione Industriale (IRI).

By the 1960s STET had become a vast, cumbersome conglomerate with numerous branches and activities. It owned five separate telecoms companies: SIP, the largest, operated the local telephone network; Iritel was the long-distance business; Italcable served international traffic; Telespazio managed satellite communications; and SIRM specialised in maritime communications.

This complex structure was eventually simplified in 1994 when Telecom Italia was formed. The five state-owned businesses were merged to create a single entity, responsible for every aspect of communications. The restructuring eliminated most duplication, hugely improved efficiency and gave Italy, for the first time, a fully functioning telecommunications system.

Despite the chronic failings of the fixed-line company, the first mobile service launched in Italy, in 1985, wasn't well received: four years later there were just 66,070 subscribers. The problem was that SIP had elected to use the non-standard home-grown RTMS for the service. It rectified this in 1990, launching a second analogue system based on the UK's TACS standard. Less than a year later, SIP had connected its 300,000th subscriber, three quarters of whom were signed up to the new service.

GSM followed in 1992, and had a few thousand users before the government banned SIP from promoting it. The government had intended to award a second mobile licence to a new entrant, but hadn't got around to it, and it was thought that were SIP allowed to continue promoting the new service, it might achieve an unassailable lead.

In 1995 it managed to get the marketing ban lifted, and changed the name of the mobile company to Telecom Italia Mobile (TIM). By the time the second entrant launched in 1995, the GSM base stood at 170,000, but, more importantly, TIM had connected over 3 million analogue customers, all of whom it hoped to eventually convert to GSM. That vast base, plus the $470-million licence fee Omnitel had had to pay for the second licence, gave TIM something of an advantage, which it would exploit to the full.

By this time, the company had already made its first investments outside Italy. In 1990 it had invested in Telecom Argentina, one half of the old monopoly PTT, Empresa Nacional de Telecomunicaciones. During 1992 STET joined up with NYNEX to apply for one of the two new mobile licences in Italy. It paid what was then the rather eye-watering sum of $160 million for a 20-year licence, but the new business failed to produce the kind of returns it was hoping for. Telecom Italia would hate to admit it, but it was comprehensively outmanoeuvred by Vodafone's Panafon consortium.

Like many of its European counterparts, Telecom Italia had plans to create a large European presence ahead of the impending deregulation. In 1996 it took a second step in this direction when it bought a 19.61% interest in Bouygues Decaux Telecom (BDT), a company controlled by the Bouygues conglomerate, which owned 55% of France's third operator, Bouygues Telecom. It followed this later in the year by acquiring 25% of Mobilkom Austria and, at the same time, 25% of Mobilkom's parent company, Telekom Austria.

Telecom Italia made one final investment in Europe during the 1990s, buying an initial 18.6% interest in the Spanish mobile operator Auna in 1997. (It subsequently raised this to 26.9%, before selling it in 2001.)

In 1995 Telecom Italia had obtained a 50% stake in Bolivian operator Entel Bolivia. Telecom Italia had outbid MCI and Telefónica in an auction for the stake, paying a full ten times the book value and, at $610 million, more than twice MCI's losing bid of $303 million.

In Brazil, Telecom Italia won the auction for control of two regional companies, Tele Celular Sul and Tele Nordeste Celular. It was also successful in its attempt to acquire new licences in other regions, including the large market of Minas Gerais, and another region where Telefónica had the wireline licence. Later, in another round of licensing, it would be able to achieve national coverage.

Towards the end of 1999 the Turkish government announced that it would issue two new GSM licences in 2000. At the time there were just over 8.1 million mobile subscribers in the country, out of a population of 62 million, and that total was split very inequitably between Turk Telekom's NMT-450 network, with 121,000 subscribers, and the two GSM operators, Turkcell and Telsim. More than half of the digital subscribers were connected to Turkcell, which had an impressive customer base of over 5.4 million.

Telecom Italia came up with an ingenious plan. The government had stated that it would award the first licence to the highest bidder; the second licence, unusually, would be awarded to any other bidder who chose to better that high bid by $10 million. The Italians realised that if they bid an extravagant amount, no one else would be prepared to match their number, let alone better it, so they would have acquired a licence that was one of three, rather than one of four. So, in partnership with the local Is Bank, they bid a massive $2,515 million. The plan worked – they won and no one came close to matching them.

Then it all went horribly wrong. The Turkish government announced that in the absence of a second bidder, they'd require the state PTT, Turk Telekom, to pay the same amount for the fourth licence. $2.5bn from one arm of government to another – there were going to be four operators after all. By 2004 it was clear that there really was only room for three operators in the country, and in 2004 the two new businesses merged.

The BDT stake was sold back to Bouygues in 2002. The Austrian interests were soon to follow: the shares in Mobilkom were bought back by Telekom Austria, and at the same time Telecom Italia sold half of its interest in Telekom Austria itself. The remaining half followed a year later.

Turk Telekom eventually bought out TIM in 2006, leaving it with two main businesses, in Italy and Brazil, and a number of smaller investments, including those in Argentina and Bolivia, both of which have subsequently been sold. This change of direction occurred because Telecom Italia was itself acquired, control of the company having passed to Olivetti in 1999 before being delivered to a consortium controlled by Pirelli, the tyre company, in 2001.

Today the company appears to be suffering the consequences of too many changes in direction, resulting from far too frequent changes in the controlling shareholder. France's Vivendi is the latest to try its hand: whether it will be any more successful than Olivetti, Pirelli and all the others that have gone before remains to be seen.

Telefónica

Telefónica was incorporated in Madrid in 1924 as Compañia Telefónica Nacional de España (CTNE),[91] with the American company ITT as one of its major shareholders.

In that first year, CTNE signed a contract with the government that gave it a monopoly on telephone service, but required it to acquire the operations and assets of the numerous independents that had been formed after the launch of the first telephone service in 1877. It also had to modernise, automate and expand the entire network. It addressed this task with alacrity, introducing automatic exchanges and establishing links with Cuba, Argentina and Uruguay, as well as the Canary Islands and Mallorca.

The outbreak of civil war in 1936 led to huge damage to the network – and the country as a whole – and much of the good work was undone.

After the end of the Second World War, the company was nationalised by General Franco's administration. ITT's stock was acquired; the state retained 41% and distributed the rest to more than 700,000 individual shareholders.

Telefónica was one of the first European companies to launch a mobile service: an NMT-450 network, in 1982. When this became capacity constrained, it switched to the TACS standard in 1990. Five years later it began operating the country's first GSM system.

Telefónica enjoyed a monopoly in mobile until later that year, 1995, when Airtel entered the market. At this point mobile penetration was just 2%, half the level in Germany, which was itself one of the slowest markets in Europe.

Telefónica chose to adopt a slightly different approach to its

---

91 The company used the name CTNE up until 1988.

northern-European counterparts. For it, Latin America was the most attractive target for its diversification efforts. Some of its competitors suggested that it chose this approach because it lacked the confidence to take on companies such as BT or Deutsche Telekom, but an alternative view is that it wanted to optimise its chances of success by using any edge it might have, including its national language. As there were more than 250 million Spanish speakers in Latin America at that time, this clearly made sense.

It did a better job in the international sphere than at home. As early as 1990 Telefónica bought a 43.6% stake in the Chilean PTT Compañia de Teléfonos de Chile, following this later in the year with the acquisition of an effective 14.1% interest in Telefónica de Argentina through its participation in the Cointel consortium. The following year it took a 16% stake in Venworld, another consortium, which went on to acquire a 40% interest in CANTV, at that time the monopoly operator in Venezuela.

Two further markets followed in 1994, Colombia and Peru. In the former, Telefónica won a stake in Compañía Celular de Colombia, a company serving the western region, an area best known for its two widely exported cash crops, coffee and cocaine. In Peru, it acquired control of two regional PTTs, Compañía Peruana de Teléfonos and Entel, which served Lima and the rest of the country respectively. Telefónica merged these to form Telefónica del Peru.

So by the mid-1990s Telefónica had accumulated an extensive portfolio of international mobile properties, most of which were in Latin America. Some of these were held directly, some through Telefónica International, a 76.2%-owned subsidiary in which the Spanish state owned the remaining 23.8%.

The year 1994 was also when Telefónica made its only European investment of the 1990s, acquiring a 60% shareholding in Telemobil, a company with a licence to operate an NMT-450 system in parts of Romania. Telefónica sold the stake in 1997.

By 1996 every country in Latin America had a working cellular system of one kind or another. The total number of connections stood at 6.69 million – Brazil had 2.7 million subscribers, making it the 11th largest market in the world, followed by Mexico (1.02 million), Argentina (687,000),

Colombia (523,000), Venezuela (445,000) and Chile (320,000). To put this into perspective, the USA was at the time by far the largest market in the world, with 44 million mobile subscribers, compared to 37.2 million in the whole of Europe. Japan had the second-largest market with 18.2 million, and the UK the third with 6.8 million, just ahead of China with 6.78 million. In total, just 136 million of the world's 5.78 billion people owned a mobile phone, equivalent to 2.35% penetration.

By that time, Brazil was served by numerous regional subsidiaries of the state-controlled Telebras plus a handful of independent operators. In early 1995, the state had announced that it was planning to privatise these regional operators and also perhaps introduce further competition. Eventually several regional holding companies were created to own controlling stakes in the regional operators. Having borrowed the methodology from America, it was somewhat inevitable that these new entities would be referred to as 'Baby Bras', albeit mercifully briefly.

Telefónica teamed with its European neighbour, Portugal Telecom, in an attempt to acquire stakes in as many of the regional companies as possible, its Compañia Riograndense de Telecomunicações consortium obtaining a controlling stake in the fixed and mobile operators serving Rio Grande do Sul, and controlling stakes in the voting shares of Telesp Participações, Tele Sudeste Celular Participações and Tele Leste Celular Participações, holding companies that owned operators serving such major population centres as Sao Paolo, Rio de Janeiro and Espirito Santo.

In 1999 Telefónica added two new markets in Central America, El Salvador and Guatemala. In 2001 it completed the acquisition of Telefónia Celular del Norte, Baja Celular Mexicana, Movitel del Noroeste and Celular de Telefónia, four of the nine regional operators set up to compete with Radiomobil Dipsa, the Teléfonos de México subsidiary. This coherent and focused strategy throughout the 1990s left Telefónica with something approaching a continental footprint in Latin America.

In the last weeks of 2002 the number of mobile subscribers in Latin America edged past the 100-million mark, most of them from just two markets, Brazil and Mexico (34.9 million and 26.2 million connections, respectively), countries with 54% of the continent's total population. There was a similar concentration at operator level: almost 30 million of the 102 million were connected to networks controlled by América Móvil, with another 21.4 million being controlled by Telefónica. BellSouth, the region's third-largest operator, accounted for a further 11.5 million, well ahead of fourth-placed Telecom Italia, which had a total of 6.4 million.

Together, these four companies accounted for over two thirds of all subscribers in the region. A year later, they accounted for almost three quarters of the total 127 million connections. América Móvil had more than 40 million customers and Telefónica was closing in on 30 million. Both were well represented in the key markets of South America – Argentina, Brazil, Chile, Colombia, Peru and Venezuela – but in Mexico, América Móvil had a huge advantage, with over 23 million customers compared to Telefónica's 3.5 million.

Although Telefónica had wasted a fair amount of money in Europe buying 3G licences it didn't really want, it was in better shape financially than some of its peers, so it was in a position to strike a deal if something suitable arose. In January 2004 AT&T Wireless, the third-largest American mobile operator, announced that it was putting itself up for sale. A few days later Cingular Wireless – a joint venture between two of the Bell regional holding companies, SBC Communications (60%) and BellSouth (40%) – offered $11 a share in cash. Vodafone surprised all concerned by counter-bidding, but then Cingular raised its offer, to $15 per share in cash, snatching AT&T Wireless from the British company's grasp. To raise its share of the cash, BellSouth put its Latin American mobile businesses up for sale.

The following month, BellSouth announced that it was selling its entire Latin American mobile business to Telefónica for a total of $5.85 billion. It was a great fit for the Spanish company: an impressive 10.5 million customers spread across nine countries, six of which were new to it. The deal gave Telefónica a total of 40.8 million connections and brought it within touching distance of América Móvil's total. The Mexican company was still out ahead with 43.5 million (thanks to its dominant position in its home market and also, somewhat ironically, its 2003 acquisition of BellSouth's two Brazilian businesses) but the gap was now much smaller. Together,

América Móvil and Telefónica controlled operators that accounted for 62.5% of all the mobile customers in Latin America.

In 2005 Telefónica turned its attention back to Europe. It acquired Cesky Telecom in the Czech Republic and followed this with a tender offer for the British company O2, a deal it completed in 2006 at a cost of €23.6 billion, making it one of its largest-ever investments. At the time of the acquisition, O2 (the former BT subsidiary) was one of the largest mobile operators in Europe, with networks in the UK, Germany, Ireland and the Isle of Man.

The problem was that the balance sheet was deteriorating. For years Telefónica had been trying to keep debt under control, taking ever-longer to pay even the most trivial bill, but eventually it could go no further in this direction. Net debt hit a new peak in 2006 after the O2 deal, at €56.4 billion. The sale of a few non-core activities, combined with cost-reduction measures, managed to cut this back to €38 billion by 2008 but by the end of 2011 debt had crept back up to its previous high level. The IPO and partial disposal of O2 Germany helped lower this from €62 billion to €55 billion, but this was still far too high.

Telefónica put O2 UK up for sale and deconsolidated it, ahead of a disposal. BT thought about buying it back. Hutchison definitely wanted to buy it, to merge it with its Three network – so Telefónica seemed well set. Ahead of the impending deal, it engaged in a bit of good housekeeping, selling O2 Ireland in 2013 and O2 Czech Republic to private equity the following year.

Then, out of the blue, the UK deal evaporated. BT preferred the look of EE to O2 and set about buying that, while the regulatory and competition people had decided that while a Hutchison takeover of O2 was fine in Ireland, it wasn't acceptable in the UK. The merger was blocked, despite the clear and obvious benefits it would have brought.

Today, Telefónica is still somewhat overweight – debt now stands at €49 billion, while debt to equity is a slightly more manageable 180%. The company looks better balanced than in the past, but it still isn't out of the woods.

Telenor

The Norwegian company Telenor has the distinction of being the only major European PTT that is still under state control. At the end of 2018 the Kingdom of Norway held 53.97% of the equity and it doesn't seem much inclined to dispose of it. It's not hard to see why: Telenor has been spectacularly successful over the past 30 years and it shows no signs of deviating from that course now.

Like most of the former European PTTs, the company that eventually became Telenor dates back to the 1870s, immediately after the invention of the telephone. The state-owned Norwegian Telegraph Administration (NTA) laid the first phone lines in the country in 1878, and two years later an offshoot of America's Bell Telephone Company established the country's first commercial operation, opening a network in Oslo. International service followed three years later.

In these early years, the various phone systems were all private enterprises, except the NTA. In 1901, the state passed legislation giving it a monopoly on the provision of telephone service in Norway, but a number of privately owned business were allowed to continue operating independently. The NTA bought out the last of these as recently as 1974.

Before that, in 1969, the NTA had changed its name to Televerket, or Norwegian Telecom, changing this again, in 1995, to Telenor, shortly after it had become a public corporation.

Telenor has been in the forefront of mobile communications for more than half a century. As the NTA, it launched the first non-cellular car phone in 1966. In 1981 it launched the NMT-450 cellular system, within a few weeks of a similar NMT network in Sweden, the first such networks in Europe and among the first in the world.

Telenor was the slowest of all the Scandinavian companies when it came to international expansion. It took its first step when it teamed up with KPN and its three Scandinavian neighbours, Telia, TDC and Telecom Finland. Together, the five formed Pannon GSM in Hungary in 1993. Two years later it acquired a licence in Montenegro for a mobile business called ProMonte GSM. This began as a consortium, through which Telenor had a 40% economic interest, but is now a wholly owned subsidiary. Three years later it took a 45% interest, in collaboration with KPN, in Ireland's second mobile operator, Esat Digiphone.

After this, in quick succession, Telenor completed two deals in Europe and a third in Asia. In the Ukraine, Telenor became a founder shareholder of Kyivstar, with an initial 35% interest. This was the third mobile operator to be licensed in the country, but it soon caught up with both its rivals. By the end of 2001 it had become the market leader, with 1.9 million connections and Telenor had secured a majority of the equity.

In Greece, Telenor took a 30% stake in Cosmote, also the third operator in the country. Its partner was OTE, the national PTT.

The third licence that year was in Bangladesh, where mobile penetration was zero, and so too was average income, or so it seemed to those who passed on the opportunity. But the partnership, with the local Grameen Bank, grew and prospered, and today the business has over 72 million customers.

Telenor had acquired minority interests in a few Russian mobile businesses over the years, but none was particularly significant. The largest of these, North-West GSM, had ended 1999 with just 40,844 subscribers in St Petersburg and the surrounding region. Then, in 1998, Telenor announced a new deal in Russia, on an altogether different scale: for NOK1.24 billion ($164 million) it bought a 31.6% stake in VimpelCom, one of largest operators in the country, with more than 350,000 customers, including in Moscow. It was the start of a long and tempestuous relationship.

The year 1999 saw a return to Western Europe, where Telenor became a 10% shareholder in BT's German venture VIAG InterKom, if only for a few months. It also joined Orange (the British company, pre-acquisition) in Austria, where the two acquired 17.45% interests in the country's third mobile carrier, optimistically branded One. And it invested in additional businesses in Russia, and in Malaysia. (The Malaysian company, now known as Digi.Com Berhad, had begun life as Mutiara, launching a GSM-1800 network in 1995; the market was difficult and Telenor was brought in to help.)

By the end of 2002 Telenor was active in 11 international markets, with the 15th-largest mobile customer base in the world.

Unlike many European operators, Telenor was not unduly troubled by the 3G auctions, thanks in large part to the exercise of two put options (in Ireland and Germany) that allowed it to sell assets of questionable value to BT at high prices, while at the same time avoiding potentially crippling licence fees.

It decided to focus on Scandinavia, certain Eastern European countries and a number of developing Asian markets. As a first step, Telenor won one of two new licences in Pakistan in 2004 and launched a national network early the following year. Next, it acquired Vodafone's Swedish subsidiary, in early 2006, following that later in the year with Mobi63 in Serbia, the country's second-largest mobile operator. Its customer base now exceeded 115 million worldwide – 25 times the population of its home market.

Over the last 25 years, Telenor has rarely put a foot wrong, but it did just that in 2008 when it acquired a 60% interest in Unitech Wireless, an Indian company that had won spectrum in 21 of India's 22 telecoms regions, at a cost of $1.2 billion. Telenor spent a huge amount of time and effort over the next few years building up a customer base of more than 50 million people, only to discover the licences had been revoked because of irregularities in the bidding process.

In the first half of 2013 Telenor took advantage of OTE's difficult financial situation and acquired its Bulgarian mobile subsidiary CosmoBul.

The following year two new licences were on offer in Myanmar, the last remaining mobile monopoly of any size. The incumbent operator, Myanmar Post & Telecommunications, had launched an AMPS network back in 1993, and had overlaid this with D-AMPS in 1995 before moving to CDMA five years later, but the service wasn't popular. By the end of 2011 there were only 1.24 million subscribers out of a population of over 52 million.

Telenor took one of the licences, and Ooredoo, the state-controlled operator from Qatar, was awarded the other. The Qataris were first out of the blocks, and by the end of their first quarter in operation they had connected over a million new customers, compared to Telenor's 281,000. Three months later Telenor had taken a lead over Ooredoo, and by the end of 2015, the two newcomers had a combined base of 19.5 million, while

Myanmar Post & Telecommunications had 18 million. The most recent numbers from Telenor, in September 2019, show that it now has more than 21.6 million subscribers on its own.

Around 2015 the stormy relationship with VEON, the former VimpelCom, finally began coming to an end when the Norwegians started selling out, and today its holding in the Russian multinational stands at just 5.7%.

With 183 million mobile customers spread over nine markets, Telenor is clearly one of the most successful of the former European PTTs.

Telia and Sonera

The four Scandinavian PTTs were among the first operators in the world to launch cellular services. The cachet this bestowed enabled all four to be successful early investors in international markets.

Recognising that they were unlikely to be able to match the firepower of a Deutsche Telekom or a regional Bell, the four aimed at the markets of Eastern Europe and other, more remote parts of the world where they were unlikely to encounter such competition. Sweden's Telia and Telecom Finland (later Sonera) were attracted to many of the same opportunities and collaborated in several. Denmark's TDC and Telenor generally preferred to go their own way.

Telia has its origins in a company that began operating just a year after the invention of the telephone, the state-owned Kongliga Elektriska Telegraf Verket (Royal Electric Telegraph Works). 'Even back in 1885, Stockholm not only had the most telephones per capita but also the most telephones in absolute numbers of any city in the world. There were more telephones in Stockholm than there were in New York, London, or Paris. Naturally, the biggest switching exchange was also in Sweden. With its 7,000 lines and super-modern multiposition switchboard, designed by Ericsson, the exchange on Malmskillnadsgatan in Stockholm was the world's biggest in 1887.'[92]

In addition, '[i]n contrast to most other countries, the telephone utility

---

92 Telia Annual Report, 1999.

in Sweden was never protected by a legal monopoly or regulations. At the start of the 1900s, there was severe competition for customers of telephone services in Stockholm, and there was no formal monopoly for telephone service—then or later. This could be the reason that Sweden has one of the world's highest numbers of telephones per capita and relatively low telephony rates, now and throughout the history of telecommunications.'[93]

Over on the other side of the Gulf of Bothnia, local service in Finland, which began in 1877, was provided by 50 or more independent local phone companies (*puhelin*), many of which were municipally owned and/or not for profit. The state-owned Telecom Finland was unique in that it had virtually no presence in the local phone market and was principally a long-distance and mobile-phone specialist (it was given the first mobile licence), but it also was obliged to provide local telephone service in the – surprisingly few – areas of the country that were so far from anywhere that they were not consider commercially viable by any of the *puhelin*.

After the liberalisation of the Finnish phone market in 1996, Telecom Finland was given the chance of entering the local fixed market while the local operators were allowed into the international and domestic long-distance markets. It began to lose market share at a rapid rate as the local companies undercut it. After just one year, it had lost 30% of the international market and 40% of the domestic. Fighting fire with fire, it cut mobile-call rates to close to fixed levels – and local-call traffic flooded on to its network. This was the first real proof anywhere in the world that mobile could be a substitute for fixed in any and every application.

Ahead of privatisation in 1998, Telecom Finland changed its name to Sonera.

In 1990 Telia and Telecom Finland began their first collaboration, the Baltic Tele joint venture. The newly independent nations in the region all faced the same problem seen in other parts of the former Soviet Union – inadequate, obsolete telecoms technology in insufficient quantity. Estonia looked west for help and Baltic Tele obliged, taking a 49% stake in a newly formed company, Eesti Mobil Telefon (which it subsequently swapped for

---

93 Ibid.

a stake in Eesti Telekom, the parent company). Three years later, the same opportunity arose in Latvia, and once again Baltic Tele took a 49% interest, this time in Latvijas Mobilais Telefons. Lithuania would follow, on similar terms, towards the end of the decade.

In 1993 both companies were early investors in North-West GSM, which would eventually form one of the foundations of MegaFon, one of the country's three large mobile operators. Russia had begun licensing cellular systems in the early 1990s in a haphazard fashion, offering concessions in some of the 89 regions into which the country was divided. Telecom Finland had a rather greater appetite for these than did Telia, and by the mid-1990s it had stakes in companies in Kaliningrad, Murmansk, Pskov and the Karelia Republic (which borders Finland).

In 1993 it took a 35% stake in a new start-up in Turkey called Turkcell. This was to prove to be much the best of Sonera's many investments. The national PTT, Turk Telecom, had launched an NMT-450 network in 1986, but by the early years of the next decade, as subscriber numbers climbed towards the 100,000 mark, this was beginning to suffer from capacity constraints. The government decided to license two new GSM operators, and within a year Turkcell had equalled the analogue system's total base. Within two years, it had three times the number.

As a result of its shareholding in Turkcell, Sonera was able to participate in another company, a joint venture called Fintur with Turkcell itself and the Cukurova Group, Turkcell's other main shareholder. Soon, Fintur had acquired licences in four former Soviet republics, Azerbaijan, Georgia, Kazakhstan and Moldova, and the business continues to prosper to this day, as does Turkcell.

Up to this point, all had been going well for Sonera. Then, in 1999, it teamed up with Telefónica Móviles to bid for licences in Germany and Italy. This ended in tears and although the consortium made an attempt to launch a service (under the name Quam), Sonera's share of the cost of the licence and the venture itself took its debt to levels that made life as an independent entity impractical. It was persuaded to merge with Telia in 2002, after which it was obliged to be less adventurous.

In 2007 TeliaSonera acquired MCT Corp, a Delaware holding company that owned majority stakes in three central Asian companies, Indigo and Somoncom in Tajikistan, and Coscom in Uzbekistan. It also owned 12.25% of Roshan, a mobile operator in Afghanistan. The following year, it bought mobile businesses in Cambodia and Nepal, giving it one of the largest and most diverse portfolios of Asian businesses, with a presence in eight different countries excluding Turkey.

For the most part, these investments were pretty successful, so it came as something of a surprise when Telia announced in 2015 that it intended to dispose of all of its Eurasian activities. The company cited political instability following the Russian invasion of Crimea and the difficulty of repatriating earnings from some of the markets, among other reasons. Finding buyers took longer than expected, but eventually, by 2018, Telia had disposed of all of its interests in the region. It dismantled the Fintur joint venture, bringing to an end one of the industry's most successful collaborations.

At the same time, it started to reduce its stake in its other large associate, the Russian company MegaFon, and by the end of 2017 the investment had been entirely liquidated.

The Sonera name has been dropped and the newly renamed Telia Company is now focused on Scandinavia, to the exclusion of everywhere else – just as it was three decades ago.

Vodacom

Before the introduction of majority rule in South Africa in 1994, the new government-in-waiting decided to open up the telecoms market by awarding two licences to operate GSM cellular networks. One licence was to be awarded to the state-owned Telkom, but as that company was thought to lack the necessary expertise, a suitable partner was sought. Vodafone was selected from a short list and went on to form Vodacom, with Telkom South Africa (51%) and the Rembrandt Group (15%).

The terms of the licence obliged Vodacom to cover 70% of the country's population within five years and generate R1 billion ($300 million) in economic activity over the first decade – seemingly daunting targets that were easily met. It launched service in April 1994 and sold all of the initially available 10,000 connections over the course of its first weekend. Three months later there were 50,000 subscribers on the network, and three months after that there were a further 50,000.

Vodacom then began to look for opportunities in other markets. An agreement with Vodafone restricted these to African countries south of the equator. In 1996 Vodacom Lesotho started operating; this was an 80%-owned subsidiary, formed with a local partner. Three years later, Vodacom won another licence in Tanzania (62% owned) and this launched in 2000. Both of these were successful.

A third investment hasn't worked out so well: in 2002 Vodafone DRC was formed with a local partner, Congolese Wireless Networks, but the relationship between the two has been troubled.

Vodacom also acquired an 85% stake in Vodacom Mozambique, in collaboration with local partners that included the local phone company, Emotel.

More recently, in 2017, it bought Vodafone's 34.94% holding in Safaricom in Kenya, taking it above the equator for the first time.

The level of demand Vodacom experienced in all these markets far surpassed its expectations. Lacking many elements of basic infrastructure, subscribers in emerging markets have been prepared to spend a higher proportion of their disposable income on communications, a service that has become a partial if not complete substitute for road or rail, healthcare and financial services.

The multinationals in the developing world

The collapse of stock-market valuations after the dot.com bubble burst in 2000 brought an end to the industry's first phase of consolidation as far as most European operators were concerned. As the new century approached, increasingly companies in parts of the world other than the USA and Europe started to look beyond their borders for investment opportunities.

Of the world's emerging markets at the turn of the millennium, Africa and Russia were typical: at the end of 1999, mobile penetration in both was way below the global average of 8%. Russia had yet to reach 1%, while Africa was at 1.1%.

So this second wave of expansion was both strategic and opportunistic. In some cases, operators merely wished to secure access to international markets at a time when their home markets were being opened to competition – the same impetus that had propelled their European counterparts. For others, it was about making hay while the sun shone – establishing beachheads in foreign markets at more realistic valuations at a time when most European operators were absent from the market, distracted by the effects of competition at home and rendered temporarily impotent thanks to the devaluation of their acquisition currency.

Telekom Malaysia/Axiata and Singapore Telecom

Singapore Telecom (SingTel) and Telekom Malaysia had been vying with each other for some years to see who could secure the best assets, not just in South East Asia but farther afield as well. As the old century gave way to the new, the intensity of this rivalry stepped up a notch or two. The deals got bigger – and the risks and rewards greater.

SingTel had been the first to venture into the international market, in 1992, taking a 16.7% stake in the NetCom consortium in partnership with Kinnevik and Ameritech. The team was later awarded the second mobile licence in Norway.

Its next moves were closer to home: in 1993, SingTel bought a 38% stake in Globe Telecom, the second mobile operator in the Philippines. The next year it made a small investment in the Thai Shinawatra Group, through which it gained an economic interest in AIS Thailand, a publicly quoted mobile operator controlled by Shinawatra. (SingTel subsequently, in 1998, bought a direct stake in AIS.)

Late in 1995 SingTel and Ameritech reprised their Norwegian partnership when they formed the ADSB Consortium with TDC of Denmark to bid for a 49.9% stake in Belgacom.

As a general rule, SingTel preferred to take large minority shareholdings rather than taking control. The one exception to this was in Australia, when it acquired 100% of Optus, the second fixed and mobile operator in the country.

Its last deal of any size was the acquisition of a 44% stake in Pacific Bangladesh Telecom, in the final quarter of 2007. This business ceased trading in 2017. However, SingTel's overall portfolio of international assets has continued to expand, thanks to the gradual increase in its stake in Bharti Airtel (now 39.5%), and Bharti's own acquisition activities, which have seen it move into Africa.

Two years after SingTel's first international acquisition, in 1994, Telekom Malaysia followed it on to the international stage, completing a number of deals over the next three years. Early in 1995 it bought a 23% stake in the Indonesian operator Excelcomindo from NYNEX; later that year it took a 60% controlling stake in Telekom Networks Malawi, the national phone company in the East African state.

After this it moved back closer to home. In 1995 Telekom Malaysia formed Dialog, a new mobile operator in Sri Lanka, and followed this with an investment in Spice Communications, an Indian company that had won licences to operate in two of the country's regions, Karnataka and Punjab. The following year, it acquired its third and last asset there when it was awarded a licence to operate a new mobile system in Bangladesh. That company, now called Robi Axiata, began operating towards the end of 1997, the year in which Telekom Malaysia teamed up with SBC to acquire a 30% stake in South Africa's Telkom.

SingTel responded to these developments by moving into Telekom Malaysia's two largest international markets, making bigger investments in both. In 2000 it bought an initial 15.5% stake in what would soon become India's largest mobile operator, Bharti Airtel. The following year SingTel bought 35% of the shares in Telkomsel, Telekom Indonesia's mobile subsidiary.

Eventually, in 2008, Telekom Malaysia spun off its mobile and international mobile assets in a separate listed company, the Axiata Group. By that time, it had sold its stakes in Telekom Networks Malawi and Telkom in South Africa but had launched a new network in Cambodia. This company, known as Hello, acquired two further networks in the country over the next few years, including Applifone, a company that had previously been owned by Telia. Four years later, the two would complete another deal, as Axiata acquired the Swedish company's 80% stake in N-Cell in Nepal.

In 2014 Axiata's publicly listed Indonesian subsidiary, renamed XL Axiata, acquired Axis, a smaller domestic competitor, from STC, the Saudi PTT. Two years later it merged its operations in Bangladesh with those of Warid, a business that had previously been bought by Bharti Airtel.

Today, both Axiata and Singapore Telecom are among the largest mobile companies in the world.

The African multinationals

Africa had a great need for modern communications, as the existing fixed infrastructure was aging and unreliable where it existed at all. Yet the first mobile networks on the continent were greeted with something close to disdain. Tunisie Telecom was the first to launch the new service, in 1985, but by the end of 1987 it had just 224 subscribers. It took over five years to connect the 1,000th line – equivalent at that stage to 0.012% penetration.

Africa's second network was opened by Telkom in South Africa during the first half of 1986 and didn't fare much better. It reached that same 1,000-line milestone by the end of the year, but progress thereafter was painfully slow, with just 400 new customers in 1987. Five years after the launch, penetration was still less than 0.025% of the country's 36-million population.

By the end of 1993 there were still fewer than 45,000 mobile connections in the whole of Africa. By that time, new networks had been launched in Egypt (1987), Morocco and the DRC (1988), Algeria and Mauritius (1989), Botswana (1991), Cote d'Ivoire, Ghana, Nigeria and Senegal (1992), and Kenya and Tanzania (1993). All these networks were based on analogue technologies, so the early lack of enthusiasm for mobile can be attributed in part to the limited capacity of the systems and high charges for both service and equipment. However, there was another factor too: the vast majority of the citizens of these countries had never owned a telephone, were unfamiliar with its utility and didn't initially recognise its potential.

By the end of 1996 there were still fewer than 1.5 million mobile subscribers. But then the arrival of digital technology reduced the cost of the service and began to alter the way mobile was perceived, and by the end of the century, just three years later, there were nearly five times as many connections – and penetration now exceeded 1%.

Africa was beginning to attract the attention of some imaginative entrepreneurs, and several multinational telecoms groups emerged at around this time. It's in large part due to their efforts that the leap from 1% to 10% penetration took place.

The first of these was Telecel International, a company founded by Rwandan-born Miko Rwayitare in 1986. Telecel is credited with launching Africa's first mobile network in what was then Zaire (now the DRC), in 1988.[94] Telecel built on this early success and by 2000 was operating in 11 African countries.

At this point, it attracted the attention of Naguib Sawiris, the chairman of Orascom Telecom. This Egyptian company had been founded in 1950, and by the end of the millennium it was one of Egypt's largest conglomerates, with interests in transportation, construction, tourism and IT. In 1996 the telecommunications arm was formally incorporated in a separate company, Orascom Telecommunications Holdings. The following year it won its first mobile licence: a 51% stake, in collaboration with France Telecom and Motorola, in MobiNil, a company controlled by the Egyptian state. The new business began operating the country's first GSM system in May 1988 and expanded rapidly.

In 1999 Orascom acquired a controlling stake in Fastlink in Jordan; six months later it was awarded a new licence in Yemen and also bought a 39% stake in Motorola's Mobilink network in Pakistan.

This was when it made its boldest move up to this point, acquiring an 80% interest in Telecel International, which took it into Benin, Burkina Faso, Burundi, the Central African Republic, the Republic of the Congo, Cote d'Ivoire, Gabon, Togo, Uganda, Zambia and Zimbabwe – 15 different countries in less than two years; quite an achievement.

In 2001 Orascom was awarded one of two BOT franchises in Syria, and it bought out Motorola, thereby increasing its stake in both Jordan and Pakistan. In June it was awarded a new licence in Algeria. Tunisia followed in 2002.

By this time it had connected a total of 2.69 million subscribers, including more than a million outside Egypt. By the end of the year, these numbers had risen to 4.3 million and 2 million, respectively.

At this point, Orascom began to contract at the same pace as it had expanded. In 2002 Telecel's operations in Benin, Burundi, the Central African Republic and Gabon were sold to Etisalat's Atlantique Telecom, while Uganda and Zambia were acquired by the Gloria Trust, a company controlled by Miko Rwayitare, the founder of Telecel.

---

94 There's some debate about this – the Tunisians and Egyptians both claim prior launched dates.

Later in the year, Orascom sold its interests in Yemen and Jordan. Three other properties, in Cote d'Ivoire and the two Congo countries, followed over the next 18 months. This left Orascom with just Pakistan, the North African properties and two new businesses, in Bangladesh and Iraq. For the next two years Orascom focused on these core markets, increasing its holdings in these businesses when possible.

In mid-2005 a company Sawiris controlled called Weather Investments bought Wind, the third-largest operator in Italy. Weather paid €12.2 billion ($14.7 billion) for the business, which was by then a 100% subsidiary of the Italian power company ENEL. To facilitate the financing of this huge deal, Sawiris transferred the majority of his 56.5% interest in Orascom – 50% plus one share – to Weather.

Later in the year, Orascom gave an indication that it too was beginning to look at other opportunities outside Africa, buying a 19.3% minority stake in Hutchison Telecom International (HTIL) at a cost of £E7.49 billion ($1.3 billion). This was a new vehicle formed by Hutchison in 2004 to manage the company's interests in India, Indonesia, Vietnam and elsewhere. But the hook-up didn't last – Orascom had been interested mainly in India, and there was little reason to continue with the holding once Hutchison had sold its interests there to Vodafone – and in 2007, less than two years later, Orascom reduced its stake in the company to 16.19%. By the end of the year, that too had gone. The disposal resulted in a small net loss.

Orascom made another disposal at the same time, selling Iraqna to Zain in 2007 for $1.2 billion. This reinforced the impression that Sawiris was now focusing on anywhere but the Middle East & Africa region, and that perception was further underlined the very next month when Orascom was awarded a licence to operate a new mobile network in North Korea – the country's first and a monopoly. The new company, Koryolink, would be 75% owned by Orascom, with the state PTT holding the balance. It launched a W-CDMA network that year, attracting 5,300 subscribers in its first month. This number passed the 3-million milestone ten years later, in 2015, and moved into profit.

Then a problem arose: the government of North Korea refused to allow Orascom to repatriate any of the dividends the company was now starting to pay. In the company's 2015 Annual Report Sawiris admitted that Orascom had lost control of the business and was no longer able to direct its affairs.

A second attempt at geographical diversification fared little better. In 2008 Orascom joined forces with a number of Canadian investors to create Globalive, with the aim of acquiring spectrum in the forthcoming AWS band auctions. It won at least one lot of 2x5MHz in every region in Canada except Quebec. Canada looked like a great opportunity: mobile penetration was low – the lowest in any G20 country[95] – due in large part to the excessive costs of ownership. The market looked competitive enough, but as Orascom discovered, it is actually controlled by two de facto regional duopolies; Rogers and Telus in the western provinces and Rogers and Bell Canada in the east. More than ten years later, little has changed. Rates are still high and penetration well below average. Every new entrant who looked to change this has either been acquired or squeezed out of the market.

Orascom's venture failed because just as it was looking to build a strong national network with attractive tariffs, a couple of the larger operators announced that they were switching from CDMA to GSM/W-CDMA. The billion-dollar capex programmes they instigated effectively tied up the nation's limited supply of network-installation engineers for, well, for as long as was necessary to stymie Globalive ... and all the rest of that latest wave of new entrants.

Meanwhile, the situation in Algeria, Orascom's largest market, was deteriorating, with growing levels of hostility between the government and the company (for which various reasons, including religious differences and plain old jealousy, have been posited), and at the end of 2010 Orascom announced that the Russian multinational VimpelCom had offered to buy the company for $6.6 billion. The businesses in Bangladesh and Pakistan would be transferred to a new holding company called Global Telecom Holdings, in which VimpelCom would have a controlling stake, while those in Egypt and North Korea were to be

---

95 Founded in 1999, the G20 (or Group of Twenty) is an international forum for the governments and central bank governors from 19 countries and the EU to discuss policy pertaining to the promotion of international financial stability and other issues.

retained by Orascom. VimpelCom would take full ownership of the rest, including Wind in Italy and the disputed Algerian property Djezzy. The Algerian government was informed that VimpelCom would definitely be retaining its stake in Djezzy, the disputed Algerian property, and at the time of writing it remains a wholly owned subsidiary of the Russian multinational. It's one thing to argue with the Egyptians, quite another to take on President Putin!

Orascom's experience highlights the difficulties faced by a go-it-alone entrepreneur. In places like Africa, it helps to have the backing of a powerful government – especially one that has a sovereign wealth fund to help finance these kinds of adventures. And few sovereign wealth funds have more financial firepower than those of the Middle East. In the early 2000s several Gulf-based multinationals began to focus on Africa, including Etisalat from the UAE, Qatar Telecom (later Q-Tel, now Ooredoo) and Saudi Telecommunications. Today all three have substantial offshore businesses, but none was quite as driven as Zain from Kuwait.

At the end of 1999, after the Kuwaiti telecoms market was opened to competition, Zain devised a plan of phased international expansion, which it named 3x3x3. This had the aim of making the company a leading regional player within three years, a leading international player three years after that, and another three years on, by 2011, a global multinational.

The first part of the plan went quite well. Zain acquired an interest in Fastlink in Jordan from Orascom in 2002 and became the controlling shareholder in 2003. Later that year it was awarded the second licence in Bahrain, and became a member of one of three regional consortia licensed to provide mobile services in Iraq. Then it acquired an 85% stake in Celtel (formerly MSI International), the largest independent operator in Africa, at a cost of $3.3 billion. This deal gave Zain access to 14 African markets with a combined population of some 275 million people.

Celtel had begun its collection of licences as early as 1995, when it acquired a 22.5% shareholding in Clovergem, a mobile operator in Uganda that at the time was 35% owned by Vodafone. It went on to buy out Vodafone and the other shareholders, and expanded into a large part of the rest of Africa. It started with Zambia (1998), then moved on to the Republic of Congo and Malawi (1999), and Gabon, Sierra Leone, Chad and the DRC (2000). In the new millennium Celtel acquired licences in Burkina Faso, Sudan, Niger and Tanzania (2001), Kenya (2004) and Madagascar (2005). So after acquiring Celtel, Zain suddenly had a presence in 20 countries – the same number as France Telecom. At that stage, only Vodafone had more.

This position of regional pre-eminence was soon challenged, and by the end of 2006 a Lebanese company called Investcom had taken Zain's mantle as the most expansive organisation in the African mobile market.

This new regional superpower was, like Zain itself, the result of a series of acquisitions. In 1982, at the height of their country's civil war, two young Lebanese brothers had had a bright idea. Realising that during those difficult and dangerous days, access to good communications could be vital, they bought a number of satellite phones, selling them on at a huge profit. They decided to focus their efforts on telecommunications and in 1991, after the war had ended, launched a mobile network in Lebanon. By 1993, the two had entered the mobile market in Guinea, and the following year they incorporated their activities as Investcom Holdings.

A licence in Ghana followed in 1996, and another in Benin in 1999. In 2000, Investcom was awarded licences in Liberia and Yemen, and the next year it won the other BOT licence in Syria. The year 2003 brought a licence in Cyprus, and 2004 another in Guinea-Bissau. Two years on Investcom acquired a 55% stake in Bashair Telecom in Sudan.

The company was now quite substantial, and trading in Investcom began on the London Stock Exchange during 2005. By the end of that year, the company had just under 5 million customers across its eight markets.

In 2006, MTN of South Africa bid for the company, valuing it at $5.5 billion. The offer was accepted.

MTN had first become involved with mobile communications when a consortium of which it was a member, and which included Cable & Wireless and SBC Communications, and the South African media company Johnnic

Holdings, was awarded one of the two GSM licences in South Africa and began operating in June 1994.

In late 1998 Cable & Wireless and SBC Communications sold out, allowing Johnnic Holdings to increase its stake and a new subsidiary, MTN International, was created. It was immediately successful: that year a consortium in which it had a 50% stake was awarded a GSM licence in Uganda, and it had similar successes in Rwanda and Swaziland; in 2000 it acquired Camtel Mobile, the holder one of two GSM licences in Cameroon; and in 2001 it acquired a licence in Nigeria, a country with a population of 124 million and a telephone penetration of just 0.4%. Just two years later MTN connected its millionth subscriber in Nigeria.

Also in 2001, MTN announced a partnership with the US military conglomerate Lockheed Martin, which would provide internet access in Africa via a satellite system, and within a year Lockheed Martin was providing service in Nigeria, South Africa and Uganda (as well as in several other African countries where the group didn't have a presence).

By 2003 the MTN Group had 6.7 million connections; a year on this had risen to 9.5 million, and by 2005 it was some 14.3 million.

During the rest of 2005 MTN acquired interests in another five businesses, in Core d'Ivoire, Zambia, Botswana, the Republic of Congo and Iran, closing the year with a total of 23.2 million subscribers. In 2006 it acquired a new licence in Afghanistan, bought the Syrian BOT operation and bid for Investcom. By the end of that year the enlarged MTN had a total customer base of just over 40 million spread across 21 countries, giving it a place on the list of the world's 20 largest mobile operators.

MTN's transformation from a poor number two in South Africa to the largest multinational in Africa didn't go unnoticed. Its shares began to climb, tripling in value in the four years up to the end of 2007, when they reached a new high of ZAC12,806. This valued the business, which by then had over 60 million subscribers, at an impressive $35 billion (£17.5 billion). The results for the first quarter of 2008 showed further sustained growth, with an 11% quarter-on-quarter increase in subscribers to 68.2 million.

In May 2008 India's Bharti Airtel made a surprise bid for 51% of the company. At the time, Bharti Airtel had just reported a remarkably similar performance – a 12% increase in customers to 69 million in India, with 3 million more in Sri Lanka and Bangladesh. Were the deal to go through, the new business would have over 140 million subscribers, giving it the fifth-largest mobile customer base in the world (bettered only by Vodafone, China Mobile, China Unicom and China Telecom). It didn't happen. MTN's shareholders thought the company worth more than the $19bn Bharti had offered for the 51% stake and decline to accept. Moreover, they didn't like the form the offer had taken – some cash, but a large amount of the consideration was made up of Bharti Airtel Global Depository Receipts, which did not carry a vote.

At this point, as Bharti was considering whether to raise its offer, the water was muddied by the news that Etisalat was considering a counter bid. Although slightly smaller than the other two operators, with only 44 million customers, it had the financial support of the government of the UAE with its almost unlimited financial resources. MTN's shares continued their upward trajectory, reaching more than ZAC15,000 in May. Bharti went away to think again, and Etisalat went away altogether.

The world's stock markets plunged into the turmoil of yet another financial catastrophe in September 2008, and it looked to all intents and purposes that there'd be no further developments.

But Bharti was determined. Having seen how the market in India had grown, it was confident that the same trick could be repeated in Africa. One year on from its first bid, it came back to the table with another offer. It was complicated and the South African government didn't like the proposal, so once again, nothing came of it, and Bharti began to contemplate other options.

By early 2010 Bharti Airtel had agreed to buy most of Zain's African properties – the businesses it had inherited when it bought Celtel in 2003. The much-vaunted 3x3x3 strategy had, it seemed, been abandoned in the face of mounting debts. The Celtel assets weren't quite in the same league as MTN but the agreed $10.7-billion valuation wasn't much more than half of what it had offered for 49% of the South African company. The deal increased Bharti's African footprint to over 450 million people and its subscriber base to over 180 million.

Since then Bharti has added to its African presence through the acquisition of businesses in the Republic of the Congo and Uganda, and the business as a whole is profitable and growing. That's not so easily said of Bharti's domestic business, which was hard hit by the arrival of Reliance Jio. To help lower Bharti's debt levels, an IPO of this business is currently in progress.

First incorporated in 1976, Etisalat began as the state-owned telephone company in the UAE and was partially privatised in 1983 when the government sold 40% of its holding. This had no real impact on the way the company was run, and it wasn't until 1999 that Etisalat took its first steps on to the international stage, acquiring a minority stake in Zantel, the telephone company that served the island of Zanzibar off the Tanzanian coast.

In the early 2000s Etisalat embarked on a sustained drive to acquire international investment opportunities. In 2004 it obtained an initial 35% stake in Etihad Etisalat, a new mobile operator in Saudi Arabia (up until this point, the Saudi mobile market had been a monopoly). The following year it acquired a 23% stake in Pakistan Telecommunications Ltd (PTCL), the national PTT, which gave it management control; and a 50% stake in Atlantique Telecom, which owned several of the old Telecel companies, in Benin, Burkina Faso, the Central African Republic, Gabon, Togo and Niger.

In 2006 Atlantique launched a new mobile network in Cote d'Ivoire, branding this MOOV, a name it would later use in many of its territories. Etisalat raised its stake in the company to 70% and took management control.

That year also saw the company take a majority stake (66%) in Etisalat Misr, Egypt's third mobile operator. (Misr was another great success: it had connected over a million customers within just seven weeks of its launch.) Later in 2007 Etisalat bought a 40% stake in a new mobile operator in Nigeria, Emerging Markets Telecommunications Services (EMTS), and also launched a mobile service in Afghanistan. The combined population of these three new countries – Nigeria, Egypt and Afghanistan – was over

250 million and their addition doubled the group's total market presence to some 14 countries with a combined population of over 500 million.

At the end of 2007 Etisalat bought a 15.97% equity stake in an Indonesian mobile operator called Excelcomindo Pratama, a national GSM operator controlled by Telekom Malaysia. (It sold some of its stake in the business when the company went public in 2010 and disposed of the rest in 2012.) In 2009 Etisalat acquired Millicom's mobile network in Sri Lanka.

In 2013 Etisalat completed a very high-profile deal, acquiring a 53% stake in Maroc Telecom from Vivendi, the French media company. Vivendi had been looking to dispose of its telecommunications assets for some while, having decided to focus on media above all else. It received €3.9 billion ($5.37 billion) for the stake, which gave Vivendi a profit, but probably not as large a one as it might have liked. Maroc Telecom extended Etisalat's reach into several more countries – Morocco, Gabon, Mali, Mauritania and Burkina Faso.

In 2015, Etisalat transferred ownership of the Atlantique Telecom businesses to Maroc Telecom, making it the fourth-largest multinational on the continent, behind Etisalat itself, Bharti Africa and the overall market leader, MTN.

Russia's multinationals – VEON, MTS and MegaFon

The phone system in the USSR was in poor shape when the Soviet regime collapsed. 'Inadequate investment in public telecommunications during the Soviet era and restrictions on access to advanced Western technology [had] resulted in an underdeveloped telephone system in Russia ... At the end of 1997 the number of access lines in Russia was 18.6 lines per 100 people, approximately three times lower than in developed countries such as the United States, the United Kingdom and Japan.'[96]

Russia had fewer than 20 million fixed lines and no mobiles; its telephone-line-installation waiting list was approximately 11 million, indicating serious pent-up demand.

---

96 VimpelCom Annual Report 1997.

The other republics together had fewer than 15 million fixed lines serving their population of more than 135 million; again, there were no mobiles. The average penetration across the empire was just over 12%, though there were material differences from one republic to the next, with Ukraine the highest at 22.5% and Tajikistan the lowest at 4.9%.

There was clearly huge potential but this was a part of the world where Western operators generally trod carefully: the politics of the region was unstable and uncertain, while the economies were characterised by 'high inflation, high unemployment, high rates of business failure, high government debt relative to gross domestic product, weak currencies, declining levels of foreign trade and declining real wages and the prospect exists of widespread bankruptcies, mass unemployment and the collapse of certain sectors of these economies.'[97]

In Russia the first cellular licences had been awarded on a regional basis, in most of the 89 regions into which the country was divided at that time. These first mobile operators were required to operate NMT-450 networks which were, by that time, obsolescent if not actually obsolete. This fact hadn't been considered when the decision was taken; NMT-450 was chosen because in many regions the military were still using the 800MHz and 900MHz bands – the frequencies used in the rest of the world for mobile communications.

Happily, parts of these bands became available in 1993, and at this point further licences were awarded, for networks using AMPS and GSM respectively. A short while later, the 1.8GHz band also became available. This was also used for GSM – or PCS – services.

Several Western operators began to take stakes in some of these businesses: the Scandinavian PTTs, Deutsche Telekom, Millicom and US West, for example, all invested in a number of regional operators.

However, the Russian authorities had managed to combine two of the worst aspects of previous licensing processes – regional franchises and different technologies – and the result was that uptake was very slow to start with. By 1995, five years after the first network was launched, there

---

97 Petersburg Long Distance Share Prospectus, 1994.

were still fewer than 100,000 subscribers in a country with a population of nearly 150 million; by the end of 2000 the number was still only 3.26 million, about 2.2% of the population.

Then, suddenly, the market came to life. The number of mobile connections rose to 7.75 million in 2001, and to 17.6 million in 2002 – in just two years Russia had gone from being the world's 39th-largest market to its 14th largest. Increases in subscriber numbers in 2003 and 2004 took it to 74 million, and by 2005 it had become the world's third-largest mobile market with nearly 120 million connections.

This period of dramatic growth coincided with the emergence of three national or near-national operators – MegaFon, Mobile TeleSystems (MTS) and VimpelCom. By 2005 these three together had over 100 million customers in Russia, some seven-eighths of the total. The seven regional operators that were associated with the former monopoly Rostelecom represented the only real challenge to these three, but even together they had fewer than half the number of subscribers claimed by MegaFon, the smallest of the three giants. Competition between the three drove down prices and raised the quality of the service, at a time when GSM handsets were becoming more and more affordable.

VimpelCom was the first of the three to start operating. (It is now known as VEON, having changed its name in February 2017.) Founded in 1992, in its start-up phase its staff 'comprised of a small group of scientists who had spent their careers developing advanced technology for radio electronics and defense equipment, including the Russian antiballistic missile defense system'.[98]

VimpelCom launched an AMPS/D-AMPS network in Moscow in 1994 under the Bee Line brand. Although equipment costs and tariffs were very high by Western standards – monthly average revenue per unit (ARPU) of $277 in 1997 – by the end of its third year VimpelCom had connected over 110,000 subscribers in Moscow and a further 6,000 in the rest of the country. By 1997 VimpelCom had obtained a listing for its shares on the New York Stock Exchange, the first Russian company to do so.

---

98 VimpelCom Annual Report to Shareholders, 1996.

The company strengthened its position in early 1998 when it was awarded GSM-1800 licences to operate in four new regions in addition to Moscow. These licences, in the Central, South (or North Caucasus), Siberian and Volga regions, gave VimpelCom access to over 100 million, or some 68%, of the total population.

Not only did this provide access to some of the more densely populated parts of the country, it also gave VimpelCom the opportunity to use better technology. GSM was preferable for several reasons: operators could mix and match equipment from several different suppliers, as it was an open architecture, unlike AMPS; it was being manufactured in far greater volumes than AMPS, so prices were generally lower; and the GSM licence requirements were less onerous. (The AMPS licences typically required 10% of the region to be covered after year one, 15% after year two, 30% by year three and 75% by year six. This wasn't always commercially sensible, and occasionally VimpelCom surrendered the licence rather than waste money on providing coverage in unpopulated wastelands.)

Although the 1998 financial crisis that hit Russia[99] took some of the wind out of the industry's sails, VimpelCom recovered quickly. At the end of the year, its finances were helped by the prospect of a $162-million cash injection from Telenor, which announced that it intended to take a 31.6% interest in the business.

By the end of 1999 VimpelCom had over 350,000 subscribers in Moscow, 132,000 of whom had opted for the GSM-1800 service launched only a year earlier. The regions had grown strongly too, with a total of 145,000 connections.

To facilitate the growth in the regions, VimpelCom had created a second, separate subsidiary in 1999 to own all of its activities outside Moscow. In 2001 a second strategic investor was brought in to help fund the development of the regional business – Alfa Group, a powerful Russian investment conglomerate, paid $103 million for an initial 13% interest in VimpelCom.

---

99 In August the Russian government devalued the ruble and subsequently defaulted on certain debt repayments. Within a month the ruble had lost two thirds of its value against the US dollar as investors ran for cover.

That year the company received a licence to operate in the Northern Region, giving it a foothold in the country's second-largest market, St Petersburg and surrounding Leningrad Oblast, which had a combined population of over 6 million. The cash injection provided by the new investor helped expand GSM coverage in all five regions where VimpelCom operated, and in 2002 the company came one step nearer national coverage when it was awarded a licence to operate GSM-1800 in the Urals. (The remaining, remote Far East region of the country would come its way in 2007, when it was awarded a national licence to operate a W-CDMA 3G system.)

VimpelCom made its first move into a foreign country in 2004, when it acquired control of a network in Kazakhstan. This was the first step in what was to prove to be a very rapid expansion programme: by early 2006 it had bought a second international business in Ukraine; a third in Tajikistan; a fourth and fifth in Uzbekistan; a mobile business in Georgia; and (its largest acquisition up to this point) a 90% interest in Armentel, the national telephone company in Armenia, for $501 million.

These acquisitions cost VimpelCom $1,055 billion, but enabled it to expand its overall footprint by more than 100 million. At the end of 2006 it had over 45 million mobile customers across its seven markets, including just under 40 million in Russia.

VimpelCom has continued to grow, and has included acquiring control of the Egyptian multinational Orascom and Italy's Wind in 2010, propelling it on to the list of the world's top ten mobile companies. However, the company has also carefully pruned its portfolio of assets over the years: it disposed of a Cambodian business in early 2013, following that in 2014 with the sale of three of Orascom's properties, in Canada, the Central African Republic and Burundi. Next, in 2015, Wind was merged with Hutchison's 3 Italia, while the Orascom venture in Pakistan was bolstered by the acquisition of Warid, a competitor.

The deal with Hutchison may have propelled Wind Tre into the top spot in Italy, but VEON wasn't happy with the performance of the business or, indeed, the regulatory setup in Italy, and most recently it sold its interest to its partner, CK Hutchison. This meant it dropped back out of the top ten, but that is probably not a matter of great concern – profitability is more important.

The next of the three giant companies to enter the market was MTS. This was formed in 1993 by a consortium consisting of Moscow City Telephone Company, three other Russian telecoms operators and two German companies, Siemens and Deutsche Telekom. It began operating a GSM network in Moscow in 1994, immediately after VimpelCom had launched its AMPS network.

In 1997, after a successful debut in Moscow, MTS was awarded licences to operate GSM systems in Tver, the Kostroma region and the Komi Republic. This marked the beginning of its path towards full national coverage.

Over the next four years, MTS struck ten separate deals that gave it a presence in 43 regions, nearly half the national total. During 1999 it grew its base by over 100% to close in on VimpelCom, ending the year with 342,000 subscribers. The following year it more than tripled that to 1.19 million. By the end of 2002 it had a total of more than 6.6 million subscribers, over half of whom were outside Moscow. By the end of 2003 this had jumped to 13.4 million, a 37% market share and two million more than VimpelCom. And it had a further half million subscribers in Ukraine and a 49% share in the Belarus-MTS, which had a further 465,000 customers.

By the beginning of 2004 MTS had secured licences in all but two regions, Penza and Chechnya (and the latter had by then taken the name Chechen Republic and declared independence from Russia).

In 2004 MTS acquired a 74% interest in Uzdonrobita, the larger of two mobile operators in Uzbekistan.

Those who studied the various documents MTS filed with the SEC throughout this period will know that the company made no bones of the fact that it was operating in a risky environment. The 20-F for the year 2000 is typical, with the section devoted to 'risk factors' beginning on page 6 and ending some 12,690 words later on page 32. By 2006 this section of the filing had grown to 20,000 words, including a fair chunk relating to the acquisition of a 51% stake in a GSM business in Kyrgyzstan.

MTS paid $150 million for the stake and a further $170 million for an option to acquire the balance. Then it transpired that the organisation MTS had been dealing with didn't actually own the business. After a long and costly court battle, MTS discovered that somehow or other, the assets in question had ended up in the hands of a company called Sky Mobile, which in turn was owned by arch-rival VimpelCom ...

This little trouble, though undoubtedly vexatious at the time, was little more than a sideshow. By 2005 MTS had over 58 million customers and was generating annual revenues of $5 billion, which in turn generated over $1.1 billion in net earnings. At this point, the company was valued at just under $28 billion. These numbers are even more impressive than VimpelCom's.

That year MTS acquired Barash Communications Technology, the largest mobile operator in Turkmenistan. Two years later, it bought K-Telecom, a similarly well-positioned company in Armenia.

As penetration began to rise in all of these markets, MTS began to consider its options. In 2009 it bought Comstar, a Russian fixed-line operator. Two years later it took control of MGTS, the Moscow City Telephone Company that had been one of its original founder-investors. This greatly strengthened its presence in the capital.

Since then, the company has increasingly looked outside the mobile-telecoms industry. It began to acquire regional ISPs and cable-TV companies, then took a stake in a Russian bank to allow it to provide mobile financial services through MTS Money. In 2014 it bought a minority stake in OZON, a Russian e-commerce business, and then bought out the company that supplied its billing system, the NVision Group.

Today, its targets are 'fintech, Internet of Things, big data, systems integration, OTT, e-health, e-sport, cybersecurity, clouds, and e-commerce'.[100]

The third company that emerged from Russia at this time was MegaFon, whose story began in 1994 when two telecom operators in St Petersburg

---

100 MTS Disclosure Presentation, 21 March 2017.

formed a joint venture company called Telecominvest. Two years after its inception, Commerzbank became a shareholder through a company called First National Holdings, and three years later Telia took a 30% stake in First National, which by then had an 85% interest in Telecominvest.

Telecominvest acquired stakes in several new GSM operators and also took stakes in fixed-line and internet companies. The first of these was North-West GSM, a company that held a licence to operate a GSM system in Russia's North-West region. The largest and most important market here was, of course, St Petersburg and the surrounding region, but North-West also operated in several other important cities, including Archangelsk, Murmansk and Kaliningrad. (North-West GSM company was originally backed by Sonera (23.5%) and Telia (12.74%) in the days before the two merged; Telecominvest had 45% and Northwest Telecom, the regional PTT, 15%.)

In 1998 Telecominvest acquired a controlling stake in a GSM operator in the Volga region, Volzhsky GSM. During the course of 2000 it added three further super-regions, the North Caucasus, Siberia and the Far East; and in 2001 it was able to provide full national coverage thanks to the acquisitions of Mobikom-Center, Uralskyi GSM and, most importantly, Sonic Duo, a joint venture between Central Telegraph Mobile and Sonera, which had a GSM licence for Moscow.

For the first time Telecominvest could compete on equal terms against MTS and VimpelCom. The newly enlarged business adopted MegaFon as its national brand, and in 2002 it merged all of its various regional operations into North-West GSM, renaming it OJSC MegaFon. By the end of that year the enlarged company was operational in five of the seven super-regions and had connected more than three million customers.

Service was launched in Siberia in 2003, and at the end of that year MegaFon had more than doubled its subscriber base to 6.36 million, which equated to about 17% of the market. The company launched its last regional network, in Russia's Far East, in 2004, and also made its first move into international markets, acquiring a 75% stake in Tajikistan's TT Mobile. MegaFon's customer base doubled in 2004 to 13.65 million, and by 2005 this had risen to 22.8 million. Total revenues in that year hit $2.39 billion, which produced an after-tax profit of $394 million – not quite on a par with its larger competitors but it was closing the gap.

Over the next few years MegaFon focused on increasing its market share, slowly chipping away at the lead VimpelCom and MTS had established. By the middle of 2007 it accounted for more than 20% of the national subscriber base, having connected its 30-millionth customer. By the end of 2009 this total had increased to just over 50 million.

In 2013 it acquired a company called Yota, which owned spectrum that had been designated for worldwide interoperability for microwave access (WiMAX) services, but it had become clear that this proposed alternative standard wasn't going to fly, so MegaFon obtained approval to use it for LTE, the new 4G standard.

Perhaps more significantly, 2013 also saw the launch of MegaFon's mobile financial-services product. Three years later it acquired voting control of Mail.ru, the largest social network in Russia. In a presentation to investors, MegaFon outlined its vision for the digital future, whereby mobile connectivity became the facilitator for a wide variety of value-added businesses, including e-commerce, education, TV and media, IoT and finance – in many respects a vision of the future very similar to that outlined by MTS.

At the time of writing, MegaFon is in the process of becoming a private company and has been delisted from the Moscow Stock Exchange. It has formed an alliance with Rostelecom, the former state-controlled PTT, to develop 5G services in Russia. Some of the new businesses, such as MegaFon TV and MegaFon Media, also appear to be gaining ground.

An alliance with the Chinese e-commerce giant Alibaba that was concluded in 2018 has now resulted in a new joint-venture e-commerce initiative, AliExpress. It's early days, but this has the potential to add an entirely new dimension to the company.

APPENDIX 3

Industry timeline

Date	Place	Event
6th century BCE	Persian Empire	Cyrus the Great establishes a postal system to help unify his vast empire.
5th century BCE	Persian Empire	Pigeons are used to carry messages.
	Greece	First heliograph and hydraulic semaphore devices invented.
200 BCE–200 CE (Han Dynasty)	China	Development of flag-based messaging system.
1st century	Roman Empire	Roman glassmakers discover that heated glass can be drawn out into flexible strands.
1684	England	Robert Hooke conceives the idea of transmitting sound over stretched wires.
1750	Scotland	'CM' suggests an electrical messaging system.
1794	France	Claude Chappe invents an optical semaphore system.
1795	Italy	Francisco Salva produces an electrical messaging device.
1799	Italy	Alessandro Volta invents the first battery, which he names the voltaic pile.

19th century

Date	Place	Event
1816	England	Francis Ronalds creates a single wire electrostatic 'telegraph' cable.
1820	Denmark	Hans Christian Oersted discovers the link between electricity and magnetism.
1832	Russia	Pavel Lvovitch Schilling creates a new single-needle design of telegraph.
1837	England	William Fothergill Cooke and Charles Wheatstone patent a design for an electrical telegraph.
1838	United States	Samuel Morse begins to develop an electrical telegraph.
1840	France	First demonstration of light refraction.
1840s–1850s		Widespread deployment of telegraph cables.

Date	Place	Event
1844	United States	Samuel Morse sends the first message over an electrical telegraph of his own design.
1845	England	Electric Telegraph Company (ETC) established by William Fothergill Cooke.
1845	England	First public telegraph line opened between London and Gosport.
1845	England	First use of telegraph to apprehend a criminal, John Tawell.
1850	England	British Electric Telegraph Company established using technology developed by Edward Highton.
1850	UK-Canada	First subsea telegraph cable deployed.
1851	England	Submarine Telegraph Company established.
1851	France	King Louis-Philippe nationalises the telegraph industry.
1851	United States	The New York and Mississippi Valley Printing Company is formed, the forerunner of Western Union.
1853	United States	Moses G Farmer demonstrates a duplex telegraph system.
1857	England	Following the 'Indian Mutiny', the British government calls for dedicated end-to-end subsea cable links to be established to link Britain to its Empire.
1861	Germany	Philipp Reis demonstrates a device that converts sound to electrical pulses and then transmits those down a wire. Reis calls the device a 'telephon'.
1861	United States	Rapid expansion of the telegraph network after the outbreak of Civil War.
1865	Scotland	James Clark Maxwell publishes 'A Dynamical Theory of the Electromagnetic Field', a paper that eventually leads to the development of microwave transmission.
1869	England	Nationalisation of the telegraph industry.
1869	Germany	Heinrich Hertz proves Maxwell's theory.
1870	France	Jean-Maurice-Émile Baudot demonstrates a multiplex telegraph capable of transmitting five separate messages simultaneously.
1876	United States	Alexander Graham Bell patents a device that converts sound to electrical pulses and then transmits those down a wire. Bell calls the device a 'telephone'.
1877	Brazil	Emperor Pedro ll buys two telephones at the Philadelphia Centennial Exhibition, the first foreign sale.
1877	China	First public telephone service launched.
1877	Germany	The postmaster general nationalises the telephone industry, bringing it under the control of the Post Office.

Date	Place	Event
1877	Japan	First public telephone service launched.
1877	United States	Bell Telephone Company is incorporated.
1877	United States	Western Union forms the American Speaking Telephone Company to compete with Bell.
1877	United States	First public telephone exchange enters service.
1877		Bell establishes subsidiaries in England, France, Germany, Portugal, Spain, Sweden and Switzerland.
1878	Sweden	Lars Magnus Ericsson begins modifying telephone instruments made by Bell and Siemens before working on his own designs. LM Ericsson is born.
1878	United Kingdom	David Hughes transmits 'aerial electric waves' from his home in London to a nearby receiver, the first-ever wireless transmission.
1878	United States	Thomas Edison invents an improved carbon transmitter.
1878	United States	Bell sues Western Union for patent infringement, the first of over 600 lawsuits the company will initiate or defend to protect its patent.
1879	United States	Theodore Vail becomes general manager of the Bell Telephone Company.
1880s–1900	United States	Expansion of the Bell network to near-national coverage.
1880	Malaya	First public telephone service launched.
1880	South Africa	First public telephone service launched.
1881	Russia	The government permits the first privately owned telephone companies to begin operating, before entering the business itself in 1884.
1881	United Kingdom	Henry Fawcett, the postmaster general, reverses the 1869 nationalisation decision and opens telephone industry to competition.
1881	United States	Bell Telephone Company acquires Western Electric and changes its name to American Bell Telephone Company.
1883	Scotland	An engineer at National Telephone Company patents a design for an automatic telephone exchange.
1889	France	The Société Francaise des Téléphones is nationalised.
1891	United States	Almon Strowger patents the world's first automatic telephone exchange.
1896	United Kingdom	Guglielmo Marconi demonstrates wireless transmission to the British Post Office, which agrees to help develop the technology.
1899	United Kingdom	Marconi transmits the first wireless signal across the English Channel.

Date	Place	Event
1899	United States	The Brown Telephone Company (later Sprint) is formed.

20th century

Date	Place	Event
1901	United Kingdom	Marconi transmits the first wireless signal across the Atlantic.
1903		WM Milner demonstrates that the signal sampling technology used in TDMA systems can also be applied to telephony.
1907	United States	JP Morgan takes control of Bell Telephone Company.
1910	Canada	First use of wireless telegraphy to apprehend a criminal, Harvey Crippen.
1911	United Kingdom	The telephone industry is subsumed into the General Post Office.
1914	United States	The government expropriates Marconi's US assets on the outbreak of WW1. Later, they're given to a new company, the Radio Corporation of America (RCA).
1915	United States	First coast-to-coast call, between Alexander Graham Bell in New York and Thomas Watson in San Francisco.
1915	United States	Western Electric designs a new switching technology that leads to development of crossbar exchanges.
1915		First transatlantic wireless telephone call.
1918	United States	AT&T becomes a branch of US Post Office.
1919	United States	AT&T becomes independent of US Post Office.
1924	Spain	ITT acquires several telephone companies and creates CTNE, later Telefónica.
1925	Canada	Julius Edgar Lilienfeld is awarded a patent for a transistor 22 years before its invention in America.
1925	Italy	Four regional telephone companies merge to form Società Italiana per l'Esercizio Telefonico (SIP).
1926	United States	Paul M Rainey of Western Electric creates the world's first facsimile machine.
1929	United Kingdom	Cable & Wireless is formed and becomes a state-owned company.
1931	United Kingdom	STC demonstrates the world's first microwave link, between Dover and Calais.
1934	Germany	Oskar Heil patents a 'field-effect transistor'.
1937	United Kingdom	Alec Reeves develops pulse code modulation to sample and encode telephone signals.
1940	United Kingdom	John Randall and Harry Boot create a 'cavity magnetron' at the University of Birmingham, enabling long-range radar to be successfully developed.

Date	Place	Event
1941–45	United Kingdom	At Bletchley Park, Alan Turing and Tommy Flowers develop the world's first digital computer, designed to decipher codes generated by German Enigma encryption machine.
1945	Spain	Franco nationalises the privately owned Compañia Telefónica Nacional de España (later Telefónica).
1946	United States	The Southwestern Bell division of AT&T launches the world's first car-phone service.
1947	Mexico	A consortium of investors acquires LM Ericsson's telephone business to form TelMex.
1947	United States	Engineers at Bell Labs develop the concept of a hexagonal cell architecture for a mobile-phone network.
1947	United States	Walter Brattain and HR Moore of Bell Labs demonstrate a working transistor.
1948	United Kingdom	Max Newman, a veteran of Bletchley Park, develops the world's first fully programmable electronic computer.
1950	Mexico	TelMex acquires ITT's Mexican telephone business.
1951	United Kingdom	Max Newman's team at Manchester University collaborate with the Ferranti Company to create the world's first fully programmable, commercial computer, one month before America's UNIVAC 1.
1956	United Kingdom	Tommy Flowers leads the team that creates the world's first digital telephone switch (but it's deemed uneconomic and is abandoned).
1956	United States	Bernard M Oliver and Claude Shannon are awarded a patent for a PCM-based telephone system, similar to the one invented by Alec Reeves in 1937.
1956		TAT-1, the first transatlantic submarine cable, begins operating between the UK and USA.
1957	Soviet Union	The first artificial satellite, Sputnik, is launched. The Space Race begins.
1958	United States	Jack Kilby at Texas Instruments produces the world's first integrated circuit.
1962	United States	Telstar, the world's first communication satellite, is launched.
1963	United States	MCI is set up to operate a long-distance service over a series of microwave relay stations.
1965	United Kingdom	Charles Kao and George Hockham develop the first practical optical fibres at STL, the UK research arm of ITT.

Date	Place	Event
1969	Brazil	The government acquires the telecommunication assets of the Canadian Traction, Light and Power Company, which controls over 60% of the telephone lines in the country.
1972	Mexico	TelMex is nationalised.
1972	Brazil	Telebras is formed to own controlling stakes in Brazil's numerous regional telephone companies.
1973	United States	Martin Cooper of Motorola makes the first call on a hand-portable cellular phone to his rival at Bell Labs.
1977	Italy	STET deploys the world's first optical-fibre cable in Turin.
1979	Japan	NTT launches a commercial car-phone service based on cellular technology – the world's first.
1981	Sweden	The first European cellular system is launched by Comvik, one month ahead of a similar launch by Televerket (later Telia).
1981	United Kingdom	BT is separated from the Post Office and becomes a private company in its own right.
1981	United Kingdom	Cable & Wireless is the first European telecoms operator to be privatised.
1981	Japan	SoftBank formed.
1982	United States	The FCC invites applications for non-wireline licences.
1982	United States	Craig McCaw applies for several non-wireline licences and subsequently incorporates McCaw Cellular.
1982		First mobile networks launched in Denmark, Finland, Norway and Spain.
1982	United Kingdom	Millicom takes a 15% stake in Racal's new cellular radio business.
1983	United States	LDDS, later WorldCom, is formed by Bernard Ebbers and other investors.
1983	United States	AT&T launches the first US cellular system, in Chicago.
1983	United States	GTE acquires long-distance network operated by Southern Pacific Railroad. This is later merged with US Telecom to become US Sprint.
1984	United States	AT&T is split into eight separate companies.
1984	United Kingdom	IPO of BT.
1984	United States	Metromedia buys Boston and Worcester franchises from Graphic Scanning for $48m.
1984		First mobile networks launched in Austria, Hong Kong, Indonesia, South Korea and the UAE.

Date	Place	Event
1985	United Kingdom	Racal Telecom (later Vodafone) launches the UK's first cellular system.
1985		First mobile networks launched in Canada, France, Ireland, Italy, Luxembourg, Malaysia, Netherlands, Oman, Tunisia and West Germany.
1986	United States	PacTel acquires a stake in Communications Industries.
1986	United States	McCaw acquires MCI's cellular interests.
1986		First mobile networks launched in Bahrain, Iceland, Israel, Puerto Rico, South Africa, Thailand and Turkey.
1987	United States	SBC acquires Metromedia's cellular assets.
1987	United States	IPO of McCaw Cellular.
1987		First mobile networks launched in Australia, China, Dominican Republic, Egypt, Kuwait, Morocco, New Zealand and Switzerland.
1988	United Kingdom	Racal Telecom (later Vodafone) listed in London and New York to prevent hostile takeover bid from Cable & Wireless.
1988		First mobile networks launched in Bahamas, Belgium, Cyprus, Macau, Mexico, Singapore, Taiwan and Venezuela.
1988	Argentina	BellSouth buys 35.9% of CRM, a company licensed to provide cellular service in Buenos Aires.
1988		TAT-8, the first optical transatlantic submarine cable, begins operating.
1989	United States	BT buys 22% of McCaw Cellular for $1.2bn.
1989	United States	McCaw Cellular bids for LIN Broadcasting, a company with cellular franchises in New York and Los Angeles. BellSouth counterbids, but McCaw is ultimately successful.
1989	United States	McCaw Cellular sells 13 cellular franchises to Contel.
1989		First mobile networks launched in Antigua & Barbuda, Argentina, Brunei, Chile, Costa Rica, Democratic Republic of Congo, Faroe Islands, Mauritius, Portugal and Sri Lanka.
1989		Motorola launches MicroTAC.
1990	United States	McCaw acquires 51.9% of LIN Broadcasting for $3,375m.
1990	Mexico	TelMex is privatised by consortium led by Carlos Slim and including SBC and France Telecom.
1990	Chile	Telefónica acquires 43.6% of CTC.
1990	United States	Motorola announces plans for a constellation of LEO satellites called Iridium.

Date	Place	Event
1990	United States	GTE buys Contel to create the second-largest cellular operator in the country.
1990	Sweden	Millicom merges with sister-company Kinnevik to form Millicom International Cellular.
1990	Russia	Foreign operators are invited to bid for cellular franchises.
1990		First mobile networks launched in Barbados, Cayman Islands, Croatia, Grenada, Guatemala, Guernsey, Hungary, Jamaica, Malta, Pakistan, Peru, Qatar, St Christopher & Nevis, St Lucia and St Vincent & the Grenadines.
1991	United Kingdom	Pacific Telesis sells its 25% stake in Microtel to British Aerospace.
1991	Japan	Digital Phone and Tu-Ka consortia formed to bid for regional cellular licences.
1991	Russia	Millicom and US West are awarded licences to operate cellular systems in Moscow.
1991	United Kingdom	Hutchison acquires 65% interest in Microtel.
1991	United States	Bell Atlantic buys Metro Mobile for $2.4bn.
1991	United States	BellSouth buys Graphic Scanning's cellular interests for $310m.
1991	Finland	World's first GSM network launched on a trial basis.
1991		First mobile networks launched in Anguilla, Bolivia, Czech Republic, Estonia, Latvia, Philippines, Russia, Slovakia, Slovenia, Tonga, Trinidad & Tobago and Uruguay.
1991	United States	Loral forms Globalstar, a LEO operator.
1992	Japan	Digital Phone and Tu-Ka are awarded the first of nine cellular licences.
1992	United States	Sprint acquires Centel.
1992	Japan	NTT creates NTT Mobile Communications Network.
1992	Italy	The government sells 25% of its shares in STET, the holding company for five separate telecoms businesses.
1992		Apple launches Newton PDA.
1992	India	First eight regional mobile licences awarded in four metropolitan areas.
1992	United States	BT sells it's 22% interest in McCaw Cellular to AT&T for $1.8bn.

Date	Place	Event
1992		First mobile networks launched in British Virgin Islands, Cambodia, Cuba, Gabon, Ghana, Greenland, Guyana, Laos, Lithuania, Paraguay, Poland, Senegal, Uzbekistan, Vietnam and Yemen.
1993	South Africa	The government issues a tender for two national GSM licences.
1993	Turkey	Turkcell and Telsim are awarded GSM licences in Turkey.
1993	Finland	Nokia launches the 2110 GSM handset.
1993	United Kingdom	Psion launches Series 3 Organiser.
1993	Turkey	Telecom Finland acquires a 35% stake in Turkcell, a start-up that had been awarded one of two GSM licences in Turkey.
1993	Russia	Telia, Telecom Finland and Telenor become founder shareholders of North-West GSM, a company that would later become MegaFon.
1993	United States	AT&T bids $12bn for McCaw Cellular; the deal completes in 1994.
1993	Sweden	NordicTel launches world's first commercial SMS service.
1993	United States	Pacific Telesis spins off AirTouch Communications as a separate company.
1993		First mobile networks launched in Belarus, Belize, Bulgaria, Burundi, Ecuador, El Salvador, Gambia, Greece, Guinea, Kazakhstan, Kenya, Monaco, Monserrat, Myanmar, Nicaragua, Nigeria, Romania, Suriname and Ukraine.
1994	Netherlands	IPO of 30% of the shares in KPN.
1994	United States	Bell Atlantic and NYNEX announce the proposed merger of their mobile activities.
1994	United States	AirTouch Communications and US West create a joint venture, pooling their US cellular interests.
1994	United States	AT&T completes acquisition of McCaw Cellular, renaming company AT&T Wireless.
1994	United States	IBM launches Simon, said to be the world's first smartphone.
1994	Italy	Telecom Italia is formed through the merger of five subsidiaries of STET, the government's holding company, for its various telecoms assets.
1994	United States	Andrew Wise files patent for a prepayment mechanism in the US. The concept is largely ignored in America but is widely adopted in other parts of the world.

Date	Place	Event
1994		First mobile networks launched in Angola, Azerbaijan, Cameroon, Colombia, Dominica, Fiji, Haiti, Iran, Kyrgyzstan, Madagascar, Marshall Islands, Nauru, Solomon Islands, Sudan, Tanzania, Turks & Caicos, US Virgin Islands and Vanuatu.
1995	Portugal	IPO of 26.3% of the shares in Portugal Telecom.
1995	Spain	The government sells 12% of its stake in Telefónica.
1995	United States	Bell Atlantic and NYNEX complete the merger of their mobile activities.
1995	India	RPG Cellular launches first mobile network, in Chennai.
1995	Belgium	The government sells 49.9% of RTT (later Belgacom) to ADSB Consortium for $2.47bn.
1995	Brazil	Ahead of introduction of competition, the government announces plans to split Telebras into 12 separate companies.
1995	Portugal	Portugal Telecom launches the first European prepaid mobile tariff.
1995	United States	AirTouch, Bell Atlantic, NYNEX and US West form Prime PCS to acquire PCS licences at the forthcoming 1900MHz spectrum auctions.
1995	United States	Sprint, TCI, Comcast and Cox Cable form the Wireless Co. consortium to acquire PCS licences at the forthcoming 1900MHz spectrum auctions.
1995	United States	Sprint spins off its traditional cellular business as 360° Corporation. The business is acquired by Alltel, an independent telco focused on rural markets.
1995		First mobile networks launched in Central African Republic, Republic of Congo, Cyprus, French Polynesia, Georgia, Gibraltar, India, Jordan, Lebanon, Malawi, Moldova, Namibia, New Caledonia, Reunion, Seychelles, Uganda and Zambia.
1996		Deutsche Telekom and France Telecom create Atlas, a joint venture designed to address international business market. Atlas subsequently creates Global One, with Sprint.
1996	United States	US West announces plans to acquire Continental Cablevision, the second-largest cable TV operator in the US.
1996	United States	After success of their mobile joint venture, Bell Atlantic and NYNEX announce plans for a full merger.
1996	United States	SBC Communications bids $16.5bn for Pacific Telesis; the deal closes in 1997.

Date	Place	Event
1996	Greece	IPO of OTE; private investors are offered 8% of company's shares.
1996	United States	WorldCom announces plans to acquire MFS Communications for $14.4bn in shares.
1996	Finland	Nokia launches the 9000 Communicator.
1996	Germany	IPO of Deutsche Telekom.
1996		First mobile networks launched in Albania, Armenia, Bosnia-Herzegovina, Burkina-Faso, Cook Islands, Cote d'Ivoire, Equatorial Guinea, Guadeloupe, Honduras, Isle of Man, Lesotho, Libya, Macedonia, Maldives, Mali, Martinique, Mongolia, Montenegro, Niue, Panama, Papua New Guinea, Serbia, Tajikistan and Zimbabwe.
1997	France	IPO of France Telecom.
1997	United States	Bell Atlantic and NYNEX merge, a $25.6bn deal.
1997	China	China Mobile raises $4.2bn in IPO, listing its shares on Hong Kong Stock Exchange.
1997	United States	WorldCom acquires MCI for $37bn in shares; the merged entity is renamed MCI WorldCom.
1997		First mobile networks launched in Djibouti, Mozambique, Niger, Samoa, Togo and Turkmenistan.
1998		All remaining European monopolies are opened to competition.
1998	United States	US West spins off its cable and international mobile assets into new company called MediaOne.
1998	Brazil	Telebras is split into 12 separate regional operating companies in preparation for privatisation.
1998	Switzerland	IPO of 34.5% of shares in Swisscom.
1998	Austria	Telecom Italia buys 25% of Telekom Austria from the government and also 25% of Mobilkom from Telekom Austria.
1998	United States	SBC Communications announces proposed acquisition of Ameritech; the deal closes in 1999.
1998	United States	First test call made on Iridium satellite network: Vice President Al Gore calls Gilbert Grosvenor, the great-grandson of Alexander Graham Bell.
1998	Finland	IPO of 22.3% of shares in Sonera (formerly Telecom Finland).
1998	Russia	Telenor acquires 31.6% of VimpelCom (later VEON) for NOK1.24bn.

Date	Place	Event
1998		First mobile networks launched in Botswana, Cape Verde, French Guiana, Kiribati, Liechtenstein, Rwanda and Swaziland.
1999	United Kingdom	Vodafone bids for AirTouch; Vodafone AirTouch is created.
1999	Mexico	TelMex spins off its mobile interests as América Móvil.
1999	Japan	Vodafone AirTouch starts to raise stakes in Japan Telecom and nine regional mobile operators.
1999	United Kingdom	BT acquires 40% share of Cellnet owned by Securicor Group.
1999	United Kingdom	Deutsche Telekom acquires one2one from Cable & Wireless and MediaOne for £8.4bn.
1999	United States	Iridium files for Chapter 11.
1999	United States	ICO Global Communications, another LEO operator, files for Chapter 11.
1999	United States	Verizon Communications and Vodafone AirTouch announce the formation of Verizon Wireless.
1999	United States	MCI WorldCom bids for Sprint; the proposed deal is blocked by government and eventually abandoned in 2000.
1999	Germany	France Telecom bids for Vodafone's 17.2% stake in E-Plus and offers to buy a 60.25% stake held by Veba and RWE.
1999	Germany	BellSouth exercises pre-emption rights on E-Plus shares, preventing France Telecom from obtaining control of business.
1999	Germany	KPN acquires 77.49% of E-Plus from BellSouth.
1999	United States	AT&T creates a tracking stock for AT&T Wireless, before selling 15.6% to public investors in 2000.
1999	United States	Hutchison acquires an initial 19.9% interest in VoiceStream for $248m. VoiceStream then acquires Aerial in Chicago and Omnipoint in New York to create a near-national GSM footprint.
1999	Germany	Mannesmann bids for Orange in the UK, valuing the business at €21bn.
1999	United Kingdom	Vodafone launches an unsolicited bid for Mannesmann, valuing the business at €124bn.
1999		First mobile networks launched in Ethiopia, Moldova, Nepal and Palestine.

21st century

Date	Place	Event
2000		AT&T and BT form a second Concert joint venture.

Date	Place	Event
2000	Spain	Vodafone begins acquiring controlling stake in Airtel from BT and various Spanish financial investors.
2000	United Kingdom	Vodafone's bid for Mannesmann becomes unconditional.
2000	United Kingdom	World's first auction of 3G spectrum begins. This and subsequent auctions elsewhere in Europe extract €108bn from mobile operators to the detriment of the entire European telecoms industry.
2000	Ireland	BT acquires Esat Telecom for £2bn.
2000	United Kingdom	Hutchison re-enters UK mobile market, winning A-block UMTS licence and creating the 3 Group.
2000	France	France Telecom acquires Orange from Vodafone for £25.1bn.
2000	China	IPO of China Unicom.
2000	Sweden	IPO of 29.4% of shares in Telia.
2000	United States	Bell Atlantic acquires GTE to form Verizon Communications.
2000	Japan	KDD, Japan's second-largest telecoms operator, merges with IDO.
2000	Norway	IPO of 22% of shares in Telenor.
2000	Austria	IPO of 25.8% of shares in Telekom Austria.
2000	United States	SBC and BellSouth merge their mobile interests to form Cingular Wireless.
2000	United States	Hutchison sells VoiceStream to Deutsche Telekom for $33bn; company's name is changed to T-Mobile US.
2000		First mobile networks launched in Chad, Kosovo, Liberia, Mauritania, Palau, Sierra Leone, Somalia and Syria.
2001	Germany	Telenor sells BT 10% of InterKom for £1,032m.
2001	Germany	VIAG sells BT 45% of InterKom for £4,452m.
2001	Japan	Vodafone AirTouch completes acquisition of Japan Telecom and nine regional mobile operators after series of complex transactions.
2001	United States	AT&T spins off AT&T Wireless to its shareholders.
2001	United States	AT&T spins off Liberty Media to its shareholders.
2001	Japan	KDDI is formed by merger of KDD with DDI.
2001	Sweden	Millicom sells FORA Telecom, a holding company for its Russian mobile assets, to Tele2, another Kinnevik enterprise.
2001	United States	AT&T merges AT&T Broadband, the former TCI, with Comcast.

Date	Place	Event
2001		First mobile networks launched in Mayotte and Kurdistani North Iraq.
2002	United Kingdom	BT spins off its remaining mobile assets in the UK, Germany and Ireland as mmO2.
2002	United States	Globalstar files for Chapter 11.
2002	United States	MCI WorldCom files for Chapter 11; company subsequently changes its name to MCI.
2002	Sweden	Merger of Telia with Sonera to create TeliaSonera.
2002		First mobile networks launched in Afghanistan, North Korea, Micronesia and Sao Tome & Principe.
2003	Brazil	BellSouth sells interests in Brazil to Telecom Americas, a joint venture between América Móvil, SBC and Bell Canada.
2003	Japan	Vodafone sells Japan Telecom to a private equity company.
2003		First mobile networks launched in Bhutan, Comoros, East Timor and Iraq.
2004	Belgium	Belgacom, the former RTT, is listed on the Brussels Stock Exchange.
2004	Japan	SoftBank acquires Japan Telecom.
2004	United States	BellSouth sells its remaining Latin American interests to Telefónica.
2004	United States	Vodafone bids for AT&T Wireless.
2004	United States	Cingular acquires AT&T Wireless.
2004		First mobile networks launched in Eritrea and Guinea-Bissau.
2005	Turkey	Vodafone acquires Telsim in Turkey for $4.7bn.
2005	United States	Verizon Communications announces plans to acquire MCI WorldCom; deal is completed in 2006.
2005	United States	Sprint merges with Nextel to create Sprint Nextel.
2005	United States	SBC Communications acquires AT&T for $18bn, adopts the AT&T name.
2005		First mobile network launched in Falkland Islands.
2006	Japan	Vodafone sells Vodafone KK to SoftBank.
2006	United States	AT&T, the former SBC Communications, acquires BellSouth for $86bn.
2006	Turkey	Turk Telekom acquires Telecom Italia's stake in AVEA, their mobile joint venture.
2007	India	Vodafone acquires a controlling stake in Hutchison Essar.

Date	Place	Event
2007	United States	Apple launches iPhone.
2007		First mobile networks launched in Aruba and Norfolk Island.
2008	China	China Unicom transfers its CDMA network to China Telecom.
2008	Greece	Deutsche Telekom buys 25% of OTE.
2008	United States	Verizon Wireless acquires Alltel for $28bn.
2009	United States	Brian Acton and Jan Koum found WhatsApp.
2009		First mobile network launched in Tuvalu.
2010	United Kingdom	T-Mobile UK and Orange UK merge to form Everything Everywhere (EE).
2010	India	Bharti Airtel acquires most of Zain's mobile interests in Africa.
2010	Mexico	América Móvil acquires TelMex, its former parent.
2010	Brazil	Telefónica acquires Portugal Telecom's 50% interest in Vivo Participações, after EU overrules objections from Portuguese government; business is subsequently merged with Telesp, Telefónica's fixed-line company serving Sao Paolo.
2011	United States	AT&T bids for T-Mobile US, offering $39bn; deal is blocked.
2012	Austria	América Móvil acquires a 21% stake in Telekom Austria.
2012	Germany	Vodafone acquires Kabel Deutschland for €7.7bn.
2012	Netherlands	América Móvil submits partial tender offer for KPN.
2012	United Kingdom	Vodafone acquires Cable & Wireless Worldwide in July.
2012	United States	SoftBank acquires a 20% stake in Sprint Nextel for $12.1bn, injecting a further $8bn into the struggling company.
2012		First mobile network launched in Ascension.
2013	United States	Verizon Communications acquires Vodafone's 45% stake in Verizon Wireless.
2014	Austria	América Móvil increases its stake in Telekom Austria to 50.8%, taking effective control of the business.
2014	Germany	Telefónica Deutschland acquires E-Plus, reducing the number of MNOs to three.
2014	Spain	Vodafone acquires Grupo Ono, a cable TV operator, for €7.2bn.
2014	United States	Sprint Nextel bids $24bn for T-Mobile US, abandoning offer after counter-bid from Iliad, a small French operator.

Date	Place	Event
2014		Facebook announces plans to acquire WhatsApp for $19bn.
2015	United Kingdom	BT acquires EE from Deutsche Telekom and Orange at a cost of €12.5bn.
2015	Ireland	3 Ireland acquires O2 Ireland, reducing the number of MNOs to three.
2015	United States	Verizon acquires AOL for $4.4bn.
2015	United States	AT&T acquires DirecTV in deal worth $67bn.
2015		First mobile network launched in St Helena.
2016	United Kingdom	Three UK's attempt to acquire O2 UK is blocked because this would reduce the number of MNOs to three.
2017	Italy	Hutchison buys out VEON's interest in Wind Tre.
2017	United States	Verizon Communications acquires Yahoo!.
2018	India	Vodafone India announces plans to merge with IDEA Cellular.
2018	United States	AT&T acquires Time Warner; the $106bn deal is eventually allowed after nearly two years of regulatory scrutiny.
2018		Vodafone acquires four European cable-TV businesses from Liberty Global at a cost of €18.4bn.
2018	Italy	3 Italia merges with Wind to form Wind Tre.
2019	United States	SpaceX launches the first 60 Starlink satellites.
2019	United States	Facebook announces plans for cryptocurrency called Libra.
2019	United States	Amazon reveals more details about Project Kuiper ahead of widespread deployment.

Mobile industry subscriber milestones

Date	Number of mobile subscribers	Market/Region
1986		
Jun-86	1,000,000	Global
1987		
Aug-87	1,000,000	United States
Aug-87	2,000,000	Global
1988		
Apr-88	1,000,000	Europe
May-88	3,000,000	Global
Nov-88	2,000,000	United States
Nov-88	4,000,000	Global
1989		
Apr-89	5,000,000	Global
Sep-89	2,000,000	Europe
Sep-89	3,000,000	United States
1990		
Mar-90	1,000,000	Asia Pacific
May-90	1,000,000	United Kingdom
May-90	4,000,000	United States
Aug-90	3,000,000	Europe
Sep-90	10,000,000	Global
Nov-90	5,000,000	United States
1991		
May-91	2,000,000	Asia Pacific

Date	Number of mobile subscribers	Market/Region
Jun-91	4,000,000	Europe
Aug-91	1,000,000	Japan
1992		
Feb-92	3,000,000	Asia Pacific
Apr-92	5,000,000	Europe
Jul-92	20,000,000	Global
Oct-92	10,000,000	United States
Nov-92	4,000,000	Asia Pacific
1993		
Jan-93	1,000,000	Germany
Jun-93	5,000,000	Asia Pacific
Sep-93	1,000,000	Italy
Sep-93	30,000,000	Global
Oct-93	1,000,000	Latin America
Dec-93	2,000,000	Japan
1994		
Jan-94	2,000,000	United Kingdom
Mar-94	10,000,000	Europe
Apr-94	2,000,000	Germany
May-94	40,000,000	Global
Jun-94	1,000,000	China
Aug-94	20,000,000	United States
Oct-94	2,000,000	Latin America
Oct-94	2,000,000	Italy
Oct-94	3,000,000	Japan
Oct-94	10,000,000	Asia Pacific
Oct-94	50,000,000	Global
1995		
Jan-95	1,000,000	South Korea
Jan-95	3,000,000	United Kingdom
Feb-95	4,000,000	Japan
Apr-95	2,000,000	China
Apr-95	4,000,000	United Kingdom

Date	Number of mobile subscribers	Market/Region
May-95	1,000,000	France
Jun-95	1,000,000	Middle East & Africa
Jun-95	3,000,000	Latin America
Jun-95	5,000,000	Japan
Jul-95	1,000,000	Thailand
Jul-95	3,000,000	Germany
Aug-95	1,000,000	Brazil
Aug-95	20,000,000	Europe
Sep-95	3,000,000	China
Sep-95	30,000,000	United States
Oct-95	3,000,000	Italy
Oct-95	5,000,000	United Kingdom
Nov-95	20,000,000	Asia Pacific
1996		
Jan-96	4,000,000	Latin America
Feb-96	4,000,000	China
Feb-96	4,000,000	Germany
Mar-96	2,000,000	South Korea
Mar-96	10,000,000	Japan
Apr-96	1,000,000	Spain
May-96	100,000,000	Global
Jun-96	2,000,000	Middle East & Africa
Jun-96	2,000,000	Brazil
Jun-96	5,000,000	Latin America
Jun-96	5,000,000	China
Jun-96	30,000,000	Asia Pacific
Jul-96	30,000,000	Europe
Aug-96	5,000,000	Italy
Sep-96	2,000,000	France
Sep-96	2,000,000	Spain
Sep-96	40,000,000	United States
Nov-96	1,000,000	Mexico
Nov-96	3,000,000	South Korea
Dec-96	3,000,000	Middle East & Africa

Date	Number of mobile subscribers	Market/Region
Dec-96	40,000,000	Asia Pacific
1997		
Jan-97	10,000,000	Italy
Feb-97	1,000,000	Philippines
Mar-97	1,000,000	Turkey
Mar-97	20,000,000	Japan
Mar-97	40,000,000	Europe
Apr-97	3,000,000	France
Apr-97	3,000,000	Spain
May-97	1,000,000	South Africa
May-97	1,000,000	Argentina
May-97	3,000,000	Brazil
May-97	4,000,000	South Korea
Jun-97	4,000,000	Middle East & Africa
Jun-97	10,000,000	China
Jun-97	50,000,000	Asia Pacific
Jul-97	5,000,000	Mexico
Aug-97	10,000,000	Latin America
Aug-97	50,000,000	United States
Sep-97	4,000,000	Brazil
Sep-97	4,000,000	France
Sep-97	5,000,000	South Korea
Sep-97	50,000,000	Europe
Oct-97	1,000,000	Colombia
Oct-97	4,000,000	Spain
Nov-97	5,000,000	Middle East & Africa
Nov-97	5,000,000	France
Dec-97	2,000,000	Argentina
Dec-97	200,000,000	Global
1998		
Feb-98	30,000,000	Japan
Mar-98	2,000,000	Mexico
May-98	1,000,000	Poland
May-98	2,000,000	Thailand

Date	Number of mobile subscribers	Market/Region
May-98	5,000,000	Brazil
May-98	5,000,000	Spain
May-98	10,000,000	Germany
Jun-98	2,000,000	Turkey
Jun-98	10,000,000	South Korea
Jul-98	2,000,000	South Africa
Oct-98	3,000,000	Turkey
Oct-98	20,000,000	China
Nov-98	1,000,000	India
Nov-98	1,000,000	Indonesia
Nov-98	3,000,000	Mexico
Nov-98	10,000,000	France
Nov-98	10,000,000	United Kingdom
Nov-98	20,000,000	Latin America
Nov-98	20,000,000	Italy
Dec-98	100,000,000	Asia Pacific
Dec-98	300,000,000	Global
1999		
Jan-99	2,000,000	Poland
Jan-99	100,000,000	Europe
Feb-99	2,000,000	Colombia
Feb-99	3,000,000	South Africa
Feb-99	3,000,000	Argentina
Feb-99	4,000,000	Turkey
Feb-99	40,000,000	Japan
Mar-99	4,000,000	Mexico
Apr-99	5,000,000	Turkey
May-99	2,000,000	Philippines
May-99	10,000,000	Middle East & Africa
May-99	30,000,000	China
Jun-99	10,000,000	Brazil
Jul-99	30,000,000	Latin America
Aug-99	1,000,000	Russia
Aug-99	3,000,000	Poland

Date	Number of mobile subscribers	Market/Region
Aug-99	20,000,000	South Korea
Sep-99	4,000,000	South Africa
Sep-99	400,000,000	Global
Oct-99	4,000,000	Argentina
Oct-99	10,000,000	Spain
Oct-99	20,000,000	Germany
Oct-99	20,000,000	United Kingdom
Nov-99	2,000,000	Indonesia
Nov-99	40,000,000	China
Dec-99	20,000,000	France
Dec-99	30,000,000	Italy
Dec-99	40,000,000	Latin America
2000		
Jan-00	4,000,000	Poland
Feb-00	1,000,000	Egypt
Feb-00	3,000,000	Philippines
Feb-00	500,000,000	Global
Mar-00	2,000,000	India
Mar-00	5,000,000	South Africa
Mar-00	50,000,000	China
Mar-00	50,000,000	Japan
Apr-00	5,000,000	Argentina
Apr-00	10,000,000	Turkey
Apr-00	200,000,000	Europe
May-00	2,000,000	Russia
May-00	4,000,000	Philippines
May-00	10,000,000	Mexico
May-00	30,000,000	Germany
May-00	50,000,000	Latin America
Jun-00	30,000,000	United Kingdom
Jul-00	3,000,000	Thailand
Jul-00	5,000,000	Poland
Aug-00	3,000,000	Indonesia
Aug-00	5,000,000	Philippines

Date	Number of mobile subscribers	Market/Region
Aug-00	20,000,000	Middle East & Africa
Aug-00	100,000,000	United States
Aug-00	200,000,000	Asia Pacific
Sep-00	20,000,000	Brazil
Sep-00	40,000,000	Germany
Oct-00	20,000,000	Spain
Nov-00	2,000,000	Egypt
Nov-00	3,000,000	India
Nov-00	3,000,000	Russia
Nov-00	40,000,000	Italy
Dec-00	40,000,000	United Kingdom
2001		
Feb-01	4,000,000	Thailand
Feb-01	30,000,000	France
Feb-01	50,000,000	Germany
Feb-01	300,000,000	Europe
Mar-01	1,000,000	Ukraine
Mar-01	4,000,000	Indonesia
Apr-01	1,000,000	Iran
Apr-01	4,000,000	Russia
Apr-01	30,000,000	Middle East & Africa
Apr-01	30,000,000	Spain
Apr-01	100,000,000	China
May-01	1,000,000	Vietnam
May-01	4,000,000	India
May-01	5,000,000	Thailand
Jun-01	5,000,000	Indonesia
Jul-01	3,000,000	Egypt
Jul-01	5,000,000	Russia
Sep-01	5,000,000	India
Sep-01	300,000,000	Asia Pacific
Oct-01	10,000,000	Philippines
Oct-01	20,000,000	Mexico
Nov-01	3,000,000	Colombia

Date	Number of mobile subscribers	Market/Region
Dec-01	1,000,000	Pakistan
Dec-01	10,000,000	Poland
Dec-01	40,000,000	Middle East & Africa
2002		
Jan-02	2,000,000	Iran
Mar-02	30,000,000	South Korea
Mar-02	1,000,000,000	Global
Apr-02	10,000,000	Russia
Apr-02	10,000,000	Thailand
May-02	10,000,000	South Africa
May-02	20,000,000	Turkey
May-02	30,000,000	Brazil
Jul-02	50,000,000	Middle East & Africa
Aug-02	1,000,000	Nigeria
Aug-02	3,000,000	Ukraine
Aug-02	4,000,000	Egypt
Sep-02	1,000,000	Bangladesh
Sep-02	4,000,000	Colombia
Sep-02	400,000,000	Asia Pacific
Oct-02	10,000,000	Indonesia
Oct-02	50,000,000	Italy
Nov-02	2,000,000	Ukraine
Nov-02	10,000,000	India
Nov-02	50,000,000	United Kingdom
Nov-02	100,000,000	Latin America
Dec-02	200,000,000	China
2003		
Jan-03	2,000,000	Pakistan
Feb-03	2,000,000	Vietnam
Mar-03	20,000,000	Russia
May-03	4,000,000	Ukraine
May-03	400,000,000	Europe
Jun-03	2,000,000	Nigeria
Jun-03	5,000,000	Colombia

Date	Number of mobile subscribers	Market/Region
Jul-03	5,000,000	Egypt
Jul-03	20,000,000	India
Jul-03	20,000,000	Thailand
Aug-03	40,000,000	Brazil
Sep-03	5,000,000	Ukraine
Sep-03	20,000,000	Philippines
Sep-03	30,000,000	Russia
Sep-03	500,000,000	Asia Pacific
Oct-03	3,000,000	Pakistan
Nov-03	3,000,000	Iran
Nov-03	40,000,000	France
Dec-03	30,000,000	Mexico
2004		
Jan-04	3,000,000	Nigeria
Jan-04	30,000,000	India
Feb-04	2,000,000	Bangladesh
Feb-04	20,000,000	Indonesia
Feb-04	40,000,000	Russia
Mar-04	3,000,000	Vietnam
Apr-04	4,000,000	Nigeria
Apr-04	4,000,000	Pakistan
Apr-04	50,000,000	Brazil
May-04	30,000,000	Turkey
Jun-04	50,000,000	Russia
Jul-04	5,000,000	Pakistan
Jul-04	10,000,000	Argentina
Jul-04	20,000,000	Poland
Jul-04	40,000,000	India
Aug-04	3,000,000	Bangladesh
Aug-04	5,000,000	Nigeria
Sep-04	4,000,000	Vietnam
Sep-04	10,000,000	Ukraine
Sep-04	100,000,000	Middle East & Africa
Sep-04	300,000,000	China

Date	Number of mobile subscribers	Market/Region
Oct-04	4,000,000	Iran
Dec-04	10,000,000	Colombia
Dec-04	30,000,000	Indonesia
2005		
Jan-05	4,000,000	Bangladesh
Jan-05	20,000,000	South Africa
Jan-05	500,000,000	Europe
Feb-05	5,000,000	Vietnam
Feb-05	40,000,000	Mexico
Mar-05	5,000,000	Iran
Mar-05	10,000,000	Nigeria
Mar-05	40,000,000	Spain
Mar-05	50,000,000	India
Apr-05	5,000,000	Bangladesh
Apr-05	10,000,000	Pakistan
Apr-05	30,000,000	Philippines
Jun-05	200,000,000	Latin America
Jul-05	10,000,000	Egypt
Jul-05	20,000,000	Ukraine
Jul-05	40,000,000	Indonesia
Jul-05	40,000,000	Turkey
Jul-05	2,000,000,000	Global
Aug-05	100,000,000	Russia
Oct-05	20,000,000	Argentina
Oct-05	200,000,000	United States
Nov-05	20,000,000	Colombia
Nov-05	30,000,000	Thailand
Dec-05	10,000,000	Bangladesh
Dec-05	20,000,000	Pakistan
Dec-05	30,000,000	Ukraine
2006		
Jan-06	30,000,000	South Africa
Jan-06	40,000,000	Thailand
Feb-06	30,000,000	Poland

Date	Number of mobile subscribers	Market/Region
Mar-06	10,000,000	Vietnam
Mar-06	200,000,000	Middle East & Africa
Apr-06	20,000,000	Nigeria
Apr-06	50,000,000	Indonesia
May-06	30,000,000	Pakistan
May-06	50,000,000	Mexico
May-06	400,000,000	China
Jun-06	100,000,000	India
Jul-06	10,000,000	Iran
Sep-06	40,000,000	Pakistan
Sep-06	50,000,000	Turkey
Oct-06	1,000,000	Ethiopia
Oct-06	40,000,000	Ukraine
Oct-06	1,000,000,000	Asia Pacific
Nov-06	20,000,000	Bangladesh
Nov-06	40,000,000	Philippines
Dec-06	30,000,000	Argentina
Dec-06	40,000,000	South Korea
Dec-06	300,000,000	Latin America
2007		
Jan-07	20,000,000	Vietnam
Jan-07	50,000,000	Pakistan
Jan-07	100,000,000	Brazil
Mar-07	20,000,000	Egypt
Mar-07	30,000,000	Nigeria
Mar-07	50,000,000	France
Apr-07	30,000,000	Colombia
Apr-07	300,000,000	Middle East & Africa
Jun-07	20,000,000	Iran
Jun-07	50,000,000	Ukraine
Jul-07	3,000,000,000	Global
Aug-07	50,000,000	Philippines
Sep-07	30,000,000	Bangladesh
Sep-07	40,000,000	South Africa

Date	Number of mobile subscribers	Market/Region
Sep-07	40,000,000	Poland
Sep-07	50,000,000	Thailand
Sep-07	200,000,000	India
Sep-07	500,000,000	China
Oct-07	30,000,000	Vietnam
Dec-07	40,000,000	Nigeria
Dec-07	40,000,000	Vietnam
Dec-07	100,000,000	Japan
2008		
Jan-08	30,000,000	Egypt
Feb-08	30,000,000	Iran
Feb-08	400,000,000	Middle East & Africa
Mar-08	2,000,000	Ethiopia
Mar-08	100,000,000	Indonesia
Apr-08	40,000,000	Bangladesh
Apr-08	40,000,000	Argentina
Apr-08	400,000,000	Latin America
May-08	100,000,000	Germany
Jun-08	50,000,000	Nigeria
Jul-08	40,000,000	Iran
Jul-08	50,000,000	Vietnam
Aug-08	300,000,000	India
Sep-08	50,000,000	Spain
Nov-08	500,000,000	Middle East & Africa
Dec-08	40,000,000	Egypt
Dec-08	40,000,000	Colombia
2009		
Jan-09	3,000,000	Ethiopia
Jan-09	4,000,000,000	Global
Mar-09	50,000,000	Iran
May-09	400,000,000	India
Jul-09	4,000,000	Ethiopia
Sep-09	50,000,000	Bangladesh
Sep-09	50,000,000	Egypt

Date	Number of mobile subscribers	Market/Region
Sep-09	200,000,000	Russia
Oct-09	2,000,000,000	Asia Pacific
Nov-09	500,000,000	India
Dec-09	5,000,000	Ethiopia
Dec-09	500,000,000	Latin America
2010		
Jan-10	100,000,000	Vietnam
May-10	50,000,000	Argentina
Aug-10	5,000,000,000	Global
Sep-10	50,000,000	South Korea
Sep-10	100,000,000	Pakistan
Sep-10	300,000,000	United States
Oct-10	200,000,000	Indonesia
Dec-10	200,000,000	Brazil
2011		
Mar-11	50,000,000	South Africa
May-11	10,000,000	Ethiopia
Jul-11	1,000,000	Myanmar
Dec-11	50,000,000	Poland
2012		
Jan-12	6,000,000,000	Global
Feb-12	3,000,000,000	Asia Pacific
Mar-12	1,000,000,000	China
May-12	2,000,000	Myanmar
May-12	100,000,000	Nigeria
Jun-12	50,000,000	Colombia
Jul-12	3,000,000	Myanmar
Aug-12	100,000,000	Philippines
Sep-12	4,000,000	Myanmar
Dec-12	5,000,000	Myanmar
Dec-12	20,000,000	Ethiopia
Dec-12	100,000,000	Mexico

Date	Number of mobile subscribers	Market/Region
2013		
Jan-13	1,000,000,000	Middle East & Africa
Apr-13	100,000,000	Bangladesh
Jun-13	100,000,000	Iran
Oct-13	300,000,000	Indonesia
2014		
Mar-14	100,000,000	Egypt
Sep-14	30,000,000	Ethiopia
Sep-14	7,000,000,000	Global
Oct-14	10,000,000	Myanmar
2015		
Feb-15	20,000,000	Myanmar
Aug-15	30,000,000	Myanmar
Aug-15	40,000,000	Ethiopia
2016		
Jan-16	400,000,000	United States
Jan-16	1,000,000,000	India
Mar-16	40,000,000	Myanmar
Sep-16	50,000,000	Ethiopia
Dec-16	50,000,000	Myanmar
2017		
Mar-17	4,000,000,000	Asia Pacific
Jun-17	8,000,000,000	Global

There are few notable milestones after this point. Several markets appear to have 'gone into reverse' as operators have taken a tougher line with inactive accounts, while many no longer include M2M (IoT) connections in their totals.

Privatisation in Europe

At the start of the 1980s almost all of the telecommunications operators in Europe were under state ownership. By the 1990s, however, that was beginning to change, and today only Telenor remains under state control.

After the privatisation of Cable & Wireless in 1981 and BT in 1984, eight years passed before the next European privatisation. In 1992 the Italian state sold about a quarter of the stake it held in STET, the holding company that owned the five separate companies that would later be brought together as Telecom Italia.

In 1995 the Netherlands sold a 30% interest in Koninklijke KPN, further reducing its holding to 45% later that year and then to 35% in 2000.

At around the same time, the Spanish state sold some of the shares it owned in Telefónica. This wasn't really a privatisation, as even when the Franco government nationalised it, the company had always had private-sector shareholders. The contract between the company and the state meant that the state had effective control, even though it was no more than a minority investor. Spain's Sociedad Estatal de Participaciones Patrimoniales reduced its shareholding by 12% in 1995, leaving it with just 19.85%. A second disposal reduced this to 1.9% in early 1997, and by the end of that year the company was entirely privately owned.

Also in 1997, the government of Denmark sold 49% of its stake in TDC. Later that year Ameritech acquired a 34.4% stake in TDC, and the government disposed of its residual stake, selling a further 7.2% to Ameritech and the balance to private investors.

The Portuguese government offered 26.3% of Portugal Telekom to private investors in 1995. A second sale a year later took the state's interest down to 51%, and by 1997 this holding was nearly halved, leaving just 25.1% in government hands.

In Belgium there were two stages to the privatisation process. First, in 1995, in a move that owed something to Eastern European privatisations, the government invited operators to bid for a 49.9% stake in the national telephone company, Regie des Telegraphes and Telephones. A consortium called ADSB Telecommunications was the highest bidder (BF73 billion or $2.47 billion) and was duly chosen. ADSB had three main shareholders

– Ameritech (35%), TDC (33%) and Singapore Telecom (27%) – the remaining 5% being held by Belgian investors. This 'privatisation' had the rather bizarre effect of transferring slightly more than 20% of RTT from the Belgian state to the Danish and Singaporean states, as both TDC and SingTel were still part owned by their own governments. ADSB eventually sold its interest in the business in 2004, when Belgacom shares became listed on the Brussels Stock Exchange.

In 1996 the government of Greece sold 8% of OTE to private investors. Further disposals in 1997 reduced the government holding to 80% and a year later to 65%. Sales in 1999 and 2001 reduced this to 40%.

Telekom Austria was separated from the Post Office in 1996 and privatised in the same year. This occurred when the state's Post und Telekommunikationsbeteiligungsverwaltungsgesellschaft sold 25% of the equity to Telecom Italia, which had already acquired 25% of Telekom Austria's Mobilkom mobile subsidiary. Telecom Italia increased its stake to 29.8% when the company was listed on the Vienna Stock Exchange in 2000, but offloaded 15% of this in late 2002, before finally selling out in early 2004.

France Telecom was privatised in 1997, just before the market became open to competition, when 25% of the company's shares were offered to private investors. A second tranche was sold in 1999, taking the state's interest down to 63.6%. Further disposals were made in 2004, leaving the government with less than 50% for the first time, and by the end of 2005 more than two thirds of the company's shares were in private hands.

Swisscom was privatised in 1998, when 34.5% of the company was sold to investors.

Ahead of its privatisation in 1988, in a move designed to reflect the company's growing international focus, Telecom Finland changed its name to Sonera Corporation. The Finnish government sold an initial 22.3% interest in the business at this point, and disposed of a further 20% a year later. Following Sonera's merger with Sweden's Telia in 2002, the Finnish state's stake in the enlarged business dropped to just 19%. Telia itself remained in the hands of the state until 2000, when 29.4% of the company was sold to investors. Following the Sonera merger, that dropped to 45.3%.

Telenor, the Norwegian PTT, was one of the last to be privatised. In 2000 the government offered a 22% stake in the business to private investors, this being followed by a second smaller sale in 2003, which took the government's stake down to 62.5%. A third and final tranche was made available the following year, further reducing the government's interest to 54%, which is where it stands today. Telenor remains the only European PTT where the state retains a majority stake in the business.

By the end of 2001, the process of privatisation in Western Europe was effectively complete. The competitive landscape had also been changed by the introduction of competition into every aspect of the telecoms market, including the provision of the local connection and all aspects of traffic. In every major economy, there were at least two mobile networks, though most countries had three or more operators.

Abbreviations

3G	Third generation
4G	Fourth generation
5G	Fifth generation
ADSL	Asymmetric digital subscriber line
Ameritech	American Information Technologies Corp
AMPS	Advanced Mobile Phone System
AWS	Advanced Wireless Service
BDT	Bouygues Decaux Telecom
BOT	Build-operate-transfer
Bps	bits per second
BSNL	Bharat Sanchar Nigam Limited
BT	British Telecom
BTA	basic trading area
CANTV	Compañía Anónima Nacional de Teléfonos de Venezuela
CATV	Community Antenna Television
CCI	Cellular Communications Inc.
CDMA	Code Division Multiple Access
CDMA 1X EV-DO	Code division multiple access one times radio transmission technology evolution data-optimised.
CEPT	Conférence Européenne des Administrations des Postes et des Télécommunications, or European Conference of Postal and Telecommunications Administrations
CGE	Compagnie Generale des Eaux

CMA	Cellular market area
Contel	Continental Telephone
CPE	Consumer premises equipment
CTE	Compañía de Telecomunicaciones de El Salvador
CTI	Compañia de Teléfonos del Interior
CTNE	Compañía Telefónica Nacional de España
D-AMPS	Digital Advanced Mobile Phone Service
DCS	Digital Communication System
DDI	Daini Denden Inc
DeTeMobil	Deutsche Telekom Mobilfunk
DoCoMo	Do Communications Mobile
DSL	Digital subscriber line
DRC	Democratic Republic of the Congo
DTH	Direct-to-home
EBITDA	earnings before interest, tax, depreciation and amortisation
EE	Everything Everywhere
EMTS	Emerging Markets Telecommunications Services
ETC	Electric Telegraph Company
Etisalat	Emirates Telecommunications
ETSI	European Technology Standards Institute
EU	European Union
EV	Enterprise Value
FCC	Federal Communications Commission
FTSE	Financial Times Stock Exchange
Gbps	gigabits per second
GEC	General Electric Company
GPT	GEC Plessey Telecommunications
GSM	Groupe Speciale Mobile, later Global System for Mobile
GTE	General Telephone and Electric
HSPDA	high-speed downlink packet access
HTIL	Hutchison Telecom International Limited
iDEN	Integrated Digital Enhanced Network
IoT	Internet of Things

IP	internet protocol
IPO	initial public offering
ISP	internet service provider
ITU	International Telecommunications Union
KPN	Koninklijke KPN (Royal Dutch KPN)
LAN	local area network
LDDS	Long Distance Discount Service
LEO	low Earth orbit
LTE	long-term evolution
Matav	Magyar Telekom
Mbps	megabits per second
MCI	Microwave Communications Inc
MFS	mobile financial services
MSA	metropolitan statistical area
MSCI	Morgan Stanley Capital Group International
MSI	Mobile Systems International
MTA	major trading area
MTN	Mobile Telephone Networks Holdings/Group
MTNL	Mahanagar Telephone Nigam Ltd
MTS	Mobile TeleSystems
MVNO	mobile virtual network operator
NAMTS	Nippon Automatic Mobile Telephone System
NASA	National Aeronautics and Space Administration
NMT	Nordic Mobile Telephone
NTA	Norwegian Telegraph Administration; Norwegian Telecommunications Administration
NTT	Nippon Telegraph and Telephone
NYNEX	New York and New England Exchange Company
OFCOM	Office of Communications
OFTEL	Office of Telecommunications
OJSC	Open joint stock company
OTT	Over the top
PABX	Private automatic branch exchange
PacTel	Pacific Telesis

Pbps	petabits per second
PCM	Pulse Code Modulation
PCN	Personal Communications Network
PCS	Personal Communication Service
PDA	Personal digital assistant
PDC	Personal Digital Cellular
PTB	Posts and Telecommunications Bureau
PTC	Polska Telefonia Cyfrowa
PTCL	Pakistan Telecommunications Ltd
PTT	Post, telegraph and telephone
R-Adag	Reliance-Anil Dhirubhai Ambani
RCA	Radio Corporation of America
RF	Radio frequency
RIL	Reliance Industries Limited
RSA	Rural statistical area
RTMS	Radio Telefono Mobile di Seconda generazione
SAT	Stockholms Allmanna Telefonaktiebolag (Stockholm Public Telephone Company)
SBC	Southwestern Bell
SEC	Securities and Exchange Commission
SFR	Société Francaise de Radiotéléphones
SGT	Société Générale des Téléphones
SIM	Subscriber identity module
SingTel	Singapore Telecom
SIP	Società Italiana per l'Esercizio Telefonico
SMR	Specialised Mobile Radio
STC	Standard Telephones and Cables
STC	Saudi Telecom Company
STD	Subscriber Trunk Dialling
Sprint	Southern Pacific Railroad Internal Network Telecommunications
TACS	Total Access Communication System
Tbps	terabits per second
TCI	Tele-Communications Inc

TCNZ	Telecom Corporation of New Zealand
TDC	Tele Danmark A/S
TD-SCDMA	Time division synchronous code division multiple access
TDMA	Time division multiple access
Telmex	Teléfonos de Mexico
TIM	Telecom Italia Mobile
TIW	Telesystems International Wireless
TMN	Telecomunicações Móveis Nacionais
TPSA	Telekomunikacja Polska SA
TRI	Technology Resources Industries
UAE	United Arab Emirates
UMTS	Universal Mobile Telephone Service
VDSL	Very-high-speed digital subscriber line
VoIP	Voice over internet protocol
WiMAX	Worldwide interoperability for microwave access

Glossary

Technical terms

1G First generation – now used to describe early analogue mobile-phone networks or technology.

2G Second generation – now used to describe early digital mobile-phone networks or technology.

3G Third-generation mobile technologies, such as W-CDMA, CDMA-1X EV-DO.

4G Fourth-generation mobile technology or LTE.

5G Fifth-generation mobile technology.

Block Radio spectrum is often split into blocks (eg, A Block, B Block) for regulatory or commercial purposes. Such designation may relate to the frequencies of each block, or may be arbitrarily assigned.

ADSL A technique that permits copper cable to carry voice and data traffic simultaneously.

AMPS The first American cellular telephone standard.

Analogue Describes a system in which signals or information are represented by a continuously variable physical quantity, such as a sound or a voltage. A digital system converts the analogue signal into a stream of binary digits (0s or 1s) and then organises these in such a way that they can convey complex information.

Attenuation The reduction in the strength of a signal over distance.

Base station A radio-based relay in a geographic mobile-phone cell that receives and transmits a radio signal between phone users, via a mobile switching centre.

BOT Describes a licence to build and operate a network (sometimes fixed, but more usually mobile) for a set duration, after which ownership of the assets and the business have to be transferred to the state.

BTA One of 493 geographic region in the United States, defined by the Rand McNally Corporation and used by the FCC to define a PCS licence area.

CDMA A technique that allows multiple transmitters to use the same communication channel simultaneously. Individual signals are assigned a specific code to prevent interference with other signals.

CDMA 1X EV-DO An American 3G mobile-phone standard optimised for data transmission.

Cellular radio A method of arranging radio-telephone equipment in cells to optimise the available radio spectrum (used rarely now).

CMA A metropolitan or rural statistical area in the USA.

CPE Telecommunications equipment housed at the consumer's premises, such as telephones, fax machines, data modems and PABXs.

D-AMPS An American 2G mobile-phone technology.

DCS Short-lived term to describe a GSM-1800 network.

DSL A technique that permits copper cable to carry voice and data traffic simultaneously.

DTH Industry jargon for satellite TV services.

Exchange The 'brain' of a telephone system, which directs the flow of traffic on the network, connecting customer premises equipment (such as a telephone handset) through a network of local cables (the local loop) to other customers' devices. Exchanges are connected to each other by a 'trunk' or long-distance transmission system. International 'gateway' exchanges connect one country's network with those in other territories. In a mobile system, the local loop is replaced by a series of base stations that use short-range radio signals to connect to each other and the exchange, which is referred to as an MSC or mobile switching centre.

Fibre-optics Ultra-pure glass used to transmit signals from lasers in a telephone or data-transmission network.

HSPDA An enhanced variant of the 3G W-CDMA standard.

iDEN A proprietary 2G technology developed by Motorola.

Institutional investors Investors who are acting on behalf of a third party, such as a pension fund or insurance company. The vast majority of shares are now owned by such institutions, rather than private investors.

IoT Formerly M2M or machine to machine, the Internet of Things is a network of interconnected electronic devices capable of transmitting or receiving data without human intervention.

Interconnect An agreement that regulates the terms under which two or more telephone networks are connected to each other. Such agreements enable calls that originate on one network to be terminated on another.

Internet protocol A set of rules for communication over the internet.

LAN A network that connects computers and other devices.

LEO An orbit around the Earth of an object with an altitude of 2,000 kilometres or less.

LTE A 4G mobile-phone technology.

MSA A geographic region in the USA defined by the US Office of Management and Budget used by the FCC when allocating the first cellular licences.

MTA One of 51 geographic region in the US, defined by Rand McNally and used by the FCC to define a PCS licence area.

MVNO A company that offers mobile-phone services using capacity on another operator's network.

NAMTS A 1G Japanese mobile-phone standard.

NMT A 1G-mobile phone standard developed by Ericsson.

Non-wireline A cellular-network operator that had not previously been involved in the telephone business.

OFCOM The UK's communications regulator.

OFTEL Now a part of OFCOM, the Office of Telecommunications was the UK's original telecommunications regulator.

Open-source software Code that the copyright owner allows others to change and distribute free of charge for any purpose.

OTT Services that are provided 'over the top' of conventional voice and data services, such as WhatsApp or Face Time.

PABX A small automatic telephone exchange owned or leased by a company or individual that interconnects to the main telephone network.

PCM A technique for converting analogue signals into a digital signal.

PCN UK name for a GSM-1800 mobile network.

PCS US name for a 2G mobile communication system.

PDA An electronic device with various functions, including a calculator, calendar and address book, and sometimes a telephone.

PDC A 2G Japanese mobile-phone standard that was only ever used in Japan.

PTT A state-owned entity that operated national post and telecommunication monopolies.

RSA A geographic region in the US defined by the US Office of Management and Budget used by the FCC when allocating the first cellular licences.

SIM An electronic device that's used to define the identity of the user in certain mobile network technologies.

SMR A proprietary 2G technology developed by Motorola.

STD A system where a caller can initiate a long-distance call without the intervention of an operator.

TD-SCDMA A Chinese non-standard 3G standard.

TDMA A technique that allows multiple transmitters to use the same communication channel simultaneously. Individual signals are assigned different time slots to prevent interference with other signals.

Tropospheric scatter A means of transmitting microwave signals beyond line of sight by bouncing them on to and off the troposphere.

Trunk lines The connections between separate telephone exchanges (a biological analogy).

UMTS The over-optimistic name given to some 3G mobile-phone standards.

VDSL A variant of DSL with the highest data transmission speed.

VoIP A technique for digitising a voice signal and transmitting it over the internet (eg. Skype and WhatsApp).

Wireline A company that operates a fixed, wire-based telephone network (term borrowed from the oil industry).

Financial terms

10-K Annual report filed with the American SEC by an American company that has shares that are traded on stock exchanges in America.

20-F Annual report filed with the American SEC by a non-American company that has shares that are traded on stock exchanges in America.

ARPU The revenue generated per unit or user, which allows the management of a company as well as investors to refine their analysis of the company's revenue-generation capability and growth at a per-unit level.

Call option An agreement between two or more investors allowing one or more parties to the deal to acquire some or all of another party's shareholding on a pre-determined basis.

EV Enterprise value, the total of a company's market capitalisation and net debt.

EV/EBITDA Enterprise Value divided by EBITDA, a measure of a company's value.

FTSE The FTSE or FTSE-100 is the leading index of stocks traded on the London Stock Exchange.

Goodwill The difference between the amount paid to acquire a business and that same business's net asset value.

IPO The initial sale of shares (also known as a flotation) in a corporation to the public, after which the shares are listed on one or more stock exchanges and may be freely traded by investors.

Market capitalisation The total value of all the shares in a company, calculated by multiplying the prevailing share price by the number of shares.

Merger The process by which two companies combine or consolidate; usually used to imply a slightly more benign outcome than that resulting from a takeover.

MSCI An American public company that provides investment tools including a number of market indexes.

OJSC A prefix before a Russian company's name implying that its shares can be publicly traded.

Paper Informal term for the use of shares rather than cash in a transaction.

Private equity Usually, a group of private investors who have come together for a specific purpose, such as an acquisition.

Put option An agreement between two or more investors allowing one or more parties to the deal to sell some or all of their shareholding to another investor on a pre-determined basis.

Takeover bid An attempt to acquire a controlling or complete interest in a company.

Tender offer A type of takeover bid where the bidder offers to acquire some or all of the shares in a company at a specific price between certain dates.

Tracking stock A type of publicly traded equity security issued by a company which 'tracks' the performance of a particular division of the parent company.

Unconditional offer When all parties to the offer, including external bodies such as regulators, agree to the deal.

Voting shares Shares that carry voting rights (not all shares do, which means that sometimes an investor can have effective control of a business despite only owning a minority stake).

What's in a name?

The names of some of the companies featured in this history changed once or more during the period under consideration. The following may help the casual reader keep track.

Airtel became Vodafone Airtel then Vodafone Spain.

AirTouch Communications was originally a part of Pacific Telesis (PacTel), one of the seven regional Bell companies in the USA. AirTouch merged with Vodafone to become – briefly – Vodafone AirTouch.

Alltel, an independent American telephone company providing mobile and fixed telephone services, spun off its fixed businesses, which were then merged with those of Valor Communications to form Windstream. The mobile business then acquired Western Wireless, which was in turn acquired by Verizon Wireless.

Ameritech was one of the regional Bell companies serving the Midwest. It was acquired by SBC.

AT&T split into eight separate companies. There were seven regional Bell companies – Ameritech, Bell Atlantic, BellSouth, NYNEX, Pacific Telesis, Southwestern Bell and US West – a long-distance company that retained the AT&T name, and the manufacturing company Western Electric. Following the spin-off of its mobile businesses as AT&T Wireless, the company was acquired by its former regional subsidiary, SBC Communications. SBC then adopted the AT&T name.

The Bell Telephone Company changed its name several times but eventually became AT&T.

Bell Atlantic was one of the original regional Bell companies. It merged with NYNEX and then with GTE, following which it changed its name to Verizon Communications. Its mobile activities, in which Vodafone had a 45% interest, became Verizon Wireless.

Bell South was one of the original regional Bell companies. It dropped the space between the two words to become BellSouth. It was acquired by AT&T, the former SBC.

BT emerged out of the United Telephone Company. It was known as Post

Office Telephones before becoming first British Telecom and later the BT Group.

Cable & Wireless formed Mercury Communications and merged this with three UK cable companies to form Cable & Wireless Communications. Cable & Wireless then took full ownership of the corporate and business activities of this company, selling the consumer business to ntl Inc., an American-owned UK cable-TV operator. The group then split into two, and the corporate activities were spun-off as Cable & Wireless Worldwide, while the traditional international telephone operations assumed the name Cable & Wireless Communications. Cable & Wireless Worldwide was acquired by Vodafone, and Cable & Wireless Communications by Liberty Global.

Cellnet became BT Cellnet for a brief while, before being demerged from BT. It changed its name to O2, after which it was acquired by Telefónica.

CenturyTel, an independent US operator based in Louisiana, acquired Qwest and then took the name CenturyLink.

Cingular Wireless was formed through the merger of the wireless businesses of SBC Communications and BellSouth. After the latter was acquired by AT&T (the former SBC), Cingular became a wholly-owned subsidiary of AT&T and adopted the name AT&T Mobility.

CGE, a French utility, became Vivendi Universal, an entertainment conglomerate. Most of the businesses within its COFIRA holding company were transferred to Cegetel, including SFR. Cegetel changed its name to SFR following a merger of the two businesses, while Vivendi has dropped 'Universal'. SFR later merged with France's largest cable-TV operator, Numericable, and shortly thereafter the enlarged business was acquired by Altice SA.

Digital Phone in Japan became J-Phone, before being renamed Vodafone Japan and subsequently SoftBank Mobile, following Vodafone's sale of its Japanese interests.

EE, or Everything Everywhere, was formed through the merger of T-Mobile UK and Orange UK. It was subsequently acquired by BT.

E-Plus became KPN Deutschland, before being merged with Telefónica Deutschland.

France Telecom started life as Société Générale des Téléphones, changing its name to France Télécom and later dropping the accents. It is now called Orange.

GTE, the largest of the US independent telephone companies, had several names, including General Telephone, before settling on GTE – General Telephone and Electric. It merged with Bell Atlantic to form Verizon Communications.

ITT, or International Telephone and Telegraph, has disposed of its telecoms manufacturing assets. Some of its other businesses now trade as ITT Inc.

KDDI, Japan's second-largest telecommunications company, was formed through the merger of the Japanese phone company KDD with two regional cellular operators, Daini Denden and Nippon Idou Tsushin.

KPN, Koninklijke KPN NV, is the telecommunications arm of the Dutch national PTT, formerly known as PTT Telecom.

Leap Wireless, an independent US operator that acquired several PCS licences, was bought by AT&T.

Mannesmann Mobilfunk D2 Privat became Vodafone D2 before becoming Vodafone Germany.

Marconi's Wireless Telegraphy Company created a broadcasting business that later became the British Broadcasting Corporation, today's BBC. Its manufacturing business became part of English Electric and subsequently the GEC. The original telegraph business became the 'Wireless' part of Cable & Wireless after it merged with various UK-based international cable companies. The US arm of the company was expropriated by the US government and became the Radio Corporation of America, or RCA. This business was subsequently acquired by a different General Electric, a US conglomerate that traces its origins back to Thomas Edison.

McCaw Cellular was acquired by AT&T and was later spun off, together with AT&T's other US mobile interests, as AT&T Wireless. That company was acquired by Cingular Wireless.

M-Cell, which was awarded one of the first two GSM licences in South Africa, changed its name to MTN.

MCI was incorporated as Microwave Communications Inc. It was acquired by WorldCom. The enlarged MCI WorldCom went bankrupt and was later acquired by Verizon Communications.

Mercury Communications was founded by Cable & Wireless, in conjunction with Barclays Bank and BP. It became part of Cable & Wireless Worldwide when its parent company split into two separate entities and was eventually acquired by Vodafone.

Mercury one2one was established by Cable & Wireless and Motorola. It was subsequently acquired by Deutsche Telekom and changed its name to T-Mobile UK. It merged with Orange UK to form Everything Everywhere, later rebranded as EE.

MetroPCS, an independent US operator that acquired several PCS licences, was acquired by T-Mobile.

Microtel became Orange and, later, a part of EE.

Millicom Inc. was established by the Kinnevik Group to operate a cellular network in the US. It subsequently merged with other Kinnevik assets to form Millicom International Cellular.

Mobile Telecommunications Company of Kuwait changed its name to Zain.

MT2 became Libertel and then Vodafone Libertel. It's now Vodafone Ziggo, a joint venture between Vodafone and Ziggo of the Netherlands.

NordicTel became Europolitan, then Vodafone Sweden, and finally Telenor Sweden after Vodafone's sale of the business.

NYNEX was one of the originally Bell companies. It was acquired by Bell Atlantic.

Omnipoint was acquired by VoiceStream, before the latter was bought by Deutsche Telekom and renamed T-Mobile US.

Omnitel and Pronto Italia became Omnitel Pronto Italia, then Omnitel, then Vodafone Omnitel and finally Vodafone Italy.

Orange was originally called Microtel. It retained the Orange name throughout three changes in ownership. France Telecom, which bought the company from Vodafone, took the name as its own. The original Orange UK business was merged with T-Mobile to form Everything Everywhere, later EE.

Pacific Telesis was originally one of the Bell companies. It spun off its domestic and international wireless activities as AirTouch Communications and was subsequently acquired by SBC Communications.

Panafon became Vodafone Panafon before becoming Vodafone Greece.

Qatar Telecom, the state-owned operator in the Gulf state, changed its name to Q-Tel, before changing this once more to Ooredoo.

SBC was originally called Southwestern Bell and was one of the seven regionals spun out of AT&T. It shortened its name to SBC, and later changed it to AT&T after it had acquired its former parent.

Sprint was so designated because it began life as Southern Pacific Railroad Internal Network Telecommunications. It has adopted various versions of its name over the years, including US Sprint and Sprint Nextel. Its first cellular businesses were spun off as 360° Communications; its PCS activities were named Sprint PCS before the merger with Nextel. The company is in the process of merging with T-Mobile US; whether this will occasion another name change is not yet decided.

Telecel became Vodafone Telecel and then Vodafone Portugal. It's not related to Telecel International.

Telecel International was a holding company with investments in several African mobile businesses. It was acquired by Orascom, but then split into two and sold off.

Telecom Finland, the state-owned long-distance monopoly, became Sonera at the time of its privatisation, and later a part of the merged Telia Sonera.

Telefónica has its origins in CTNE – Compañía Telefónica Nacional de España. It later became known as Telefónica de España and then Telefónica.

Televerket became Telia and, after its merger with Sonera, Telia Sonera. It has recently dropped the Sonera and reverted to Telia.

Telsim, the second mobile operator in Turkey, was acquired by Vodafone and became Vodafone Turkey.

US West was one of the seven regional Bell companies. It sold its domestic cellular interests to AirTouch and the following year created a new vehicle, US West Media Group, to own all of its unregulated activities. The telephone assets were acquired by Qwest Communications, while the Media Group was divided between AT&T, which took the US domestic assets, and Deutsche Telekom, which acquired the international operations. Qwest was in turn acquired by CenturyTel after which the enlarged business took on the name CenturyLink.

VEBA was a German electrical power utility. It changed its name to E.ON after it merged with VIAG.

Verizon Communications was formed through the merger of Bell Atlantic and GTE. Its mobile interests were merged with the American assets of the newly created Vodafone AirTouch to form Verizon Wireless. Vodafone later sold its interest in the business.

VIAG InterKom became InterKom after VIAG and Telenor sold their stakes to BT. It was known for a while as O2 Germany, before becoming Telefónica Deutschland.

VimpelCom has experimented with various versions of its name, including VimpelCom, Vimpel Communications and Vimpel-Communications. It finally gave up on all of these and changed its name to VEON Ltd.

Vodacom is a joint venture between the Vodafone Group and Telkom South Africa.

The Vodafone Group was originally founded by Racal Electronics, when it formed Racal Millicom. This became Racal Telecom after Racal bought out Millicom and Hambros Advanced Technology Trust, its original partners. Racal Telecom was the holding company for various activities – Racal Vodafone, Racal Vodac, Racal Vodata, Racal Vodapage, etc. Racal Telecom was then floated on the stock exchange as Racal Telecom Group or RTG. Following a full demerger from Racal, the business changed its name to Vodafone Group. After the merger with AirTouch, the Vodafone Group became Vodafone AirTouch for a few months, before reverting to Vodafone Group.

VoiceStream, a company created by Western Wireless, was later bought by Alltel.

Western Wireless spun off its PCS activities as VoiceStream. This then became T-Mobile US. The original cellular business was subsequently acquired by Alltel.

WorldCom was originally Long Distance Discount Service, changing its name after numerous acquisitions. After it acquired MCI, it changed its name to MCI WorldCom. MCI WorldCom was acquired by Verizon Wireless after it emerged from bankruptcy protection.

Zain was formerly known as Mobile Telecommunications Company.

Acknowledgements

To Michael Jordaan, the man who suggested I ask Benedict Evans to help with this book. To Ben Evans, the man who said, 'Sorry, I'm too busy, but perhaps you should chat to John Tysoe.'

To John, for the superhuman physical and emotional energy invested in writing this book, for a lifetime of notes taken and history recorded, and for his friendship.

To Louise, Russell and Tracey, for their blood, sweat and tears.

Most of all, to my wife, Sibella. Without her, nothing.

Alan Knott-Craig

I'd like to thank Alan Knott-Craig for providing me with the necessary motivation to write this story; Tracey Hawthorne for her excellent editing (which has left me with enough material to contemplate a second volume) and Benedict Evans, for his thoughtful agency.

Thanks are also due to those who offered me their insights at various stages over the years, most notably Sir Christopher Gent and Sir Julian Horn-Smith, but also many others, including Ken Hydon, Geoff Varrall, Ross Cormack, Stephen Davidson, Chris McFadden, Tim and Melissa Brown, Jason Crisp and Terry Barwick.

Finally, I would like to thank my four children for their continuing support and inspiration.

John Tysoe

Index

Alan Knott-Craig can be contacted at alan@herotel.com,
or visited at 156 Dorp Street, Stellenbosch, South Africa
www.bigalmanack.com

John Tysoe can be contacted at john@themobileworld.com